Transport Terminals and Modal Interchanges

Planning and Design

Transport Terminals and Modal Interchanges

Planning and Design

Christopher Blow

AMSTERDAM • BOSTON • HEIDELBERG • LONDON • NEW YORK • OXFORD
PARIS • SAN DIEGO • SAN FRANCISCO • SINGAPORE • SYDNEY • TOKYO
Architectural Press is an imprint of Elsevier

Architectural Press
An imprint of Elsevier
Linacre House, Jordan Hill, Oxford OX2 8DP
30 Corporate Drive, Burlington MA 01803

First published 2005

Copyright © 2005, Christopher Blow. All rights reserved

The right of Christopher Blow to be identified as the author of this work
has been asserted in accordance with the Copyright, Designs and
Patents Act 1988

No part of this publication may be reproduced in any material form (including
photocopying or storing in any medium by electronic means and whether
or not transiently or incidentally to some other use of this publication) without
the written permission of the copyright holder except in accordance with the
provisions of the Copyright, Designs and Patents Act 1988 or under the terms of
a licence issued by the Copyright Licensing Agency Ltd, 90 Tottenham Court Road,
London, England W1T 4LP. Applications for the copyright holder's written
permission to reproduce any part of this publication should be addressed
to the publisher

Permissions may be sought directly from Elsevier's Science and Technology Rights
Department in Oxford, UK: phone: (+44) (0) 1865 843830; fax: (+44) (0) 1865 853333;
e-mail: permissions@elsevier.co.uk. You may also complete your request on-line via the
Elsevier homepage (www.elsevier.com), by selecting 'Customer Support'
and then 'Obtaining Permissions'

Every effort has been made to contact owners of copyright material; however, the author
would be glad to hear from any copyright owners of material produced in this book
whose copyright has unwittingly been infringed.

British Library Cataloging in Publication Data
A catalogue record for this book is available from the British Library

Library of Congress Cataloging in Publication Data
A catalog record for this book is available from the Library of Congress

ISBN 0 7506 5693 X

For information on all Architectural Press publications
visit our website at www.architecturalpress.com

Typeset by Newgen Imaging Systems (P) Ltd., Chennai, India
Printed and bound in Great Britain

Working together to grow
libraries in developing countries

www.elsevier.com | www.bookaid.org | www.sabre.org

ELSEVIER   BOOK AID International   Sabre Foundation

Contents

Preface — viii
Acknowledgements — ix

Chapter 1 Introduction — 1

Chapter 2 History – landmarks in the twentieth century — 3
    2.1 Gatwick, 1936 — 3
    2.2 Other successful multi-modal interchanges (see Chapter 6) — 6

Chapter 3 The future development of integrated transport — 7
    3.1 Resolution of complexity of transport systems caused by dis-integration — 7
    3.2 Problems in resolving dis-integration — 8
    3.3 Examples — 9

Chapter 4 Two particular studies point the way — 16
    4.1 UK Highways Agency study, 'Underground Transport Interchanges' — 16
    4.2 The Scott Brownrigg/RCA InterchangeAble research programme: reclaiming the interchange — 19

Chapter 5 Twenty-first-century initiatives — 34
    5.1 International Air Rail Organisation (IARO) — 34
    5.2 International Air Transport Association (IATA) — 34
    5.3 BAA — 35
    5.4 Royal Institute of Chartered Surveyors, UK (RICS) — 36
    5.5 Chartered Institute of Logistics and Transport (CILT) — 36
    5.6 Transport for London — 37
    5.7 Nottingham University, UK — 37
    5.8 National Center for Intermodal Transportation, USA (NCIT) — 37

Chapter 6 Taxonomy of rail, bus/coach and air transport interchanges — 38
    6.1 Airport/railway interchange: vertical separation — 38
        6.1.1 Zurich Airport, Switzerland — 38
        6.1.2 Amsterdam Schiphol Airport, The Netherlands — 43
        6.1.3 Vienna Airport, Austria — 46
        6.1.4 Heathrow Airport Terminal 5, UK — 48
        6.1.5 Heathrow Airport Terminal 4, UK — 51
        6.1.6 Chicago O'Hare Airport, USA — 55
    6.2 Airport/railway interchange: contiguous — 62
        6.2.1 Paris Charles de Gaulle Airport, France — 62
        6.2.2 Frankfurt Airport, Germany — 65
        6.2.3 Stansted Airport, UK — 70
        6.2.4 Gatwick Airport, UK — 74
        6.2.5 New Hong Kong Airport at Chek Lap Kok — 76
        6.2.6 Portland International Airport, USA — 79
    6.3 Airport/railway interchange: linked adjacent — 82
        6.3.1 Lyon St Exupéry Airport, France — 82

Contents

	6.3.2	Birmingham Airport, UK	86
	6.3.3	Manchester Airport, UK	92
	6.3.4	Heathrow Airport Central Terminal Area, UK	103
	6.3.5	Southampton Airport, UK	106
	6.3.6	Atlanta Hartsfield-Jackson Airport, USA	108
	6.3.7	Ronald Reagan Washington National Airport, USA	109
	6.3.8	San Francisco International Airport, USA	111
	6.3.9	Inchon Airport, Seoul, South Korea	113
6.4	Airport/railway interchange: remote		116
	6.4.1	Luton Airport, UK	116
	6.4.2	Düsseldorf Airport, Germany	118
6.5	Multiple railway station/bus and coach/car interchanges: vertical separation		123
	6.5.1	Lyon Perrache Railway Station, France	123
	6.5.2	Circular Quay Interchange, Sydney, Australia	125
6.6	Multiple railway station/bus and coach/car interchanges: contiguous		127
	6.6.1	Ashford International Station, Kent, UK	127
	6.6.2	Channel Tunnel Terminal, Cheriton, Kent, UK	131
	6.6.3	Manchester Piccadilly Station, UK	133
	6.6.4	Stratford Station, London	136
	6.6.5	St Pancras Station, London	142
	6.6.6	Enschede Station, The Netherlands	146
	6.6.7	Rotterdam Central Station, The Netherlands	149
	6.6.8	Perth Station, Western Australia	152
6.7	Ship and ferry terminals		154
	6.7.1	Southampton Mayflower Terminal, UK	154
	6.7.2	Yokohama port terminal, Japan	157
	6.7.3	Sydney Overseas Passenger Terminal, Australia	158

Chapter 7 Common standards and requirements — 162
- 7.1 Space standards — 162
- 7.2 Security — 162
- 7.3 Border controls — 163
- 7.4 Building design legislation — 163
- 7.5 Needs of passengers with reduced mobility — 163
- 7.6 Commercial opportunities — 163
- 7.7 Terminal operator's requirements — 164
- 7.8 Transport operator's requirements — 164
- 7.9 Car parking — 164

Chapter 8 Bus and coach interface — 165
- 8.1 Vehicles — 165
- 8.2 Factors affecting size of station — 168

Chapter 9 Rail interface — 171
- 9.1 Heavy rail systems — 171
- 9.2 Light rail systems — 172

Chapter 10 Airport interface — 173
- 10.1 Airport terminal planning — 173
- 10.2 Airport terminal capacity and size — 173
- 10.3 Constraints on building form — 174
- 10.4 Overall functional planning and passenger segregation in terminals — 175
- 10.5 The aircraft interface, terminal or remote parking — 176

10.6	Landside functions	177
10.7	Airside functions	181
10.8	Aircraft and apron requirements	184

Chapter 11 Twenty-first-century trends   190
- 11.1 Evolution of transport in relation to the city   190
- 11.2 Problems affecting transportation, 2000–2025   190
- 11.3 Converging standards at the interchange   191
- 11.4 Commercial motives for development of interchanges   192
- 11.5 Reclamation of the interchange: social, commercial and sustainable   193

Bibliography   194
Index   195

Preface

With 25 years' experience of transport terminals, principally airports, I have become conscious of the real contribution that design and organisation of transport terminal buildings can make to the public transport crusade.

All too often, the designer of one terminal is forced to accept a compromise in locating and providing inter-modal links to others.

The opportunity arose in 2002 for me and my company, Scott Brownrigg Ltd – architects based in London and Guildford, UK – to sponsor research in the area of interchange design. The fruit of this research combines with a survey of airport-based and rail station- and port-based interchange facilities to point to success in the joining up of transport modes and reclaiming the interchange as a positive experience for the traveller.

My recent appointment as a Visiting Professor in the School of Engineering at the University of Surrey is affording me the opportunity to further develop integrated transport solutions.

Without good interchanges, public transport will miss the opportunity to ameliorate the travel experience of crowded roads, and the public transport usage target percentages set by local and national government and airport authorities will not be met.

Chris Blow
Guildford

Acknowledgements

I am grateful for encouragement from:

Prof. Norman Ashford and Dr Robert Caves, formerly at Loughborough University
Ted McCagg of NBBJ, Seattle
Martyn Wallwork of Mott MacDonald
Prof. Fred Lawson
Dr Bob Griffiths of the University of Surrey

And for help with initiating the research project at the RCA from Peter Forbes of Alan Stratford & Associates, and subsequent support from Prof. Jeremy Myerson, Dr John Smith and Fiona Scott in connection with the programme at the Helen Hamlyn Research Centre.

And for the support of colleagues at Scott Brownrigg Ltd, and practical help from Claire Lupton.

I acknowledge assistance from other individuals and organisations (in order of appearance in the book):

Alan Geal of Pleiade Associates
Colin Elliff of WS Atkins
John Scholey of Halliburton KBR
Andrew Sharp of IARO
Alex Bradley of IATA
Donnathea Campbell of Nicholas Grimshaw & Partners
Gilles Hondius of NACO BV
Peter Mayerhofer of Vienna Airport
Chris Chalk of Mott MacDonald
Daniel Curtin of the Department of Aviation, City of Chicago
Murphy & Jahn, Chicago (architects)
Perkins & Will, Chicago (architects)
J.-M. Chevallier and Didier Aujouannet of Aeroports de Paris
Undine Lunow of JSK International Architekten GmbH
Foster & Partners
Paula Willett of BAA Aviation Photo Library
Ann Becklund and Jennette Zarko of Trimet, Portland
Renaud Chassagne, colleague at CRB Architectes, Lyon
Richard Heard, Steve Hill, Liz Thomas and Justine Hunt at Birmingham International Airport
Bob Longworth and Ian Howarth of Manchester Airport
David Kingdom and Dawn Wadsworth of Aedas (architects)
Jonathan Rixon of The Manser Practice
Thomas Sullivan of the Metropolitan Washington Airport Authority
Catherine Murphy of Terry Farrell and Partners
Jacobs in association with Nick Derbyshire Design Associates
Charles Kevin of Eurotunnel plc
Peter Shuttleworth and Sheri Besford of BDP
CTRL Press Office
Michael Stacey of BSR Architects
James Hulme of Alsop Architects
Katie Sandoval of Foreign Office Architects
Lawrence Nield of Bligh Voller Nield, Sydney

Permission to use material has been sought wherever possible.

Photographers

Fiona Scott: Figures 4.8, 4.10, 4.12, 4.14, 4.16, 4.18, 6.136 and 6.137.
Ralph Bensberg: 6.2 and 6.3.
Hedrich/Blessing: 6.31 and 6.32.
Stefan Rebscher: 6.37 and 6.38.
Asylum Images de Synthèse: 6.60.
Renaud Chassagne: 6.61–6.63.
J.F. Marin: 6.65.
Ansgar M. van Treeck: 6.113 and 6.114.
Andreas Wiese: 6.115, 6.116, 6.118 and 6.119.
Jason Randall: 6.132.
David Barbour/ BDP: 6.133.
Author: 6.17, 6.18, 6.91–6.95, 6.124, 6.126–6.129, 6.153, 6.154 and 6.159.

1 Introduction

This book is the first of its kind to review a trend in transport systems which has been active since the advent of mass travel but has only recently come of age. The author takes the airport as the focal point of most truly multi-modal passenger terminals, and the book is therefore complementary to many books about airports and airport terminals, capturing the interest in these buildings as representing a new experience in travel.

Whether that experience is wholly satisfactory is dependent upon the exigencies of weather and congestion in an already overloaded system, as well as the way designers and managers have addressed contingency planning. Ease of transfer between different modes of public transport is a great contribution to mobility which can be made by the transport industries. This is the future to which the book points.

In many ways, the future of the movement of goods or cargo also depends upon modal transfer. The warehousing and distribution industries address the assembly and breakdown of loads between individual consumers, town centres, factories and airport cargo terminals with different degrees of containerisation. Only where passenger and goods systems can converge are they addressed in this book: an innovative idea for using rail stations as distribution points for deliveries to towns is shown in Chapter 4.1.3.

The author recognises the organic nature of modern technologically-derived buildings with a taxonomy, a classification of passenger terminal buildings as if they were bio-forms, with integral structures or linked and contiguous forms.

Examples are culled from Britain and the rest of Europe as well as North America and the Far East, where personal mobility is most concentrated. Other case studies from Australia, for example, illustrate the breadth of opportunity worldwide.

Most cities have evolved separate land-based, air and, in some cases, waterborne transport over the years, and their very separateness may have been their salvation in the face of organisation, responsibility and physical constraints. But now, with transport coming of age, that separation is being replaced by 'joined-up thinking' and transport authorities are able to exploit interchange opportunities.

Integration is often only achieved by rebuilding and replacing dis-integrated facilities. Therefore, the book captures potent examples of four-dimensional design and construction, redevelopment projects which inconvenience in the short term but rectify past wrongs in the long term. The 'grafting on' of rail systems to airports, rather than vice versa, has been achieved at Heathrow Central Area, Manchester, Schiphol and Lyon, for example.

Airport *interchange* means not just getting to the airport by public transport. It means 'choice' and synergy. A hub airport gives better choice of flights. A real interchange point gives more – it gives choice of modes.

European examples are Paris Charles de Gaulle, Zurich, Vienna, Lyon (both Aeroport St Exupéry and Perrache rail/bus station), Schiphol, Düsseldorf, Frankfurt, Manchester (both airport and Piccadilly rail station), Birmingham, Heathrow, Gatwick, Stansted, Southampton, Luton, Ashford Station, the Channel Tunnel terminal at Cheriton (rail interchange),

St Pancras and Stratford in London, and Enschede and Rotterdam in the Netherlands.

There are several examples in the USA (Chicago, Portland, Atlanta, Washington and San Francisco), but Seoul and Hong Kong are pre-eminent in the Far East and there are interesting ones in Australia, including Perth WA.

Special cases are Circular Quay in Sydney (which combines ships, ferries, rail and bus) and Yokohama in Japan.

Many common and converging standards apply to the three modes of transport operative at the interchanges with which this book is concerned: bus or coach, rail and air transport. Functional requirements and processes particular to each interface in the interchange are reviewed briefly and illustrated.

The book therefore aims to promote modal interchange as an increasingly essential feature of passenger transport terminal buildings.

2 History – landmarks in the twentieth century

Taking as a starting point the inheritance of the great railway building age of the nineteenth century and the start of reliable bus and tram operations, where did the twentieth century take us?

For the reasons highlighted in the previous chapter, not very far in the direction of 'joined-up' transport, apart from the ubiquitous bus serving the less ubiquitous railway station. For several decades, air transport had little effect, but by the time when the number of passengers in the UK, for example, using airports becomes commensurate with every single member of the population arriving or departing from an airport several times a year, that is no longer true.

2.1 Gatwick, 1936

The original satellite at London's Gatwick Airport is a generic form of circular building serving parked aircraft. A fascinating story surrounding the design of an airport terminal in 1934 is told in *Gatwick – The Evolution of an Airport* by John King. It concerns the birth of the idea of a circular terminal building by Morris Jackaman, the developer of the original Gatwick Airport in Sussex.

> One problem which particularly concerned Morris was the design of the passenger terminal. He considered that conventional terminal buildings such as Croydon, which had been described as only fit for a fifth rate Balkan state, were inefficient and not suited to expansion of passenger traffic … It is believed that one idea he considered was building the terminal over the adjoining railway. The result of his deliberations was ultimately the circular design which is a feature of the 1936 passenger terminal, now generally known as the Beehive. How this came about is intriguing. Morris was working late one night at his parents' Slough home when his father came into his study. 'Oh, for heaven's sake, go to bed,' his father urged. 'You're just thinking in circles.' Instantly Morris reacted. 'That's it, a circular terminal.' Morris quickly put his thoughts on the advantages of a circular terminal on to paper. Using the patent agents E. J. Cleveland & Co., a provisional specification was submitted to the Patent Office on 8 October 1934. Entitled 'Improvement relating to buildings particularly for Airports', the invention sought 'to provide a building adapted to the particular requirements at airports with an enhanced efficiency in operation at the airport, and in which constructional economies are afforded'.

Various advantages of a circular terminal were detailed. They included:

1 Certain risks to the movement of aircraft at airports would be obviated.
2 More aircraft, and of different sizes, could be positioned near the terminal at a given time.
3 A large frontage for the arrival and departure of aircraft would be obtained without the wastage of space on conventional buildings.

Morris's application went on to describe the terminal as 'arranged as an island on an aerodrome' and

> The building thus has what may be termed a continuous frontage and the ground appertaining to each side of it may be provided with appliances such as gangways, preferably of the telescopic sort, to extend radially for sheltered access to aircraft. It will be observed that by this arrangement the aircraft can come and go without being substantially impeded by other aircraft which may be parked opposite other sides of the building. This not only ensures efficiency of operations with a minimum delay, but also ensures to some extent at any rate that the aircraft will not, for example, in running up their engines, disturb other aircraft in the rear, or annoy the passengers or personnel thereof. In order to give access to the building without risk of accident or delay of aircraft, the building has its exit and entrance by way of a subway or subways leading from within it to some convenient point outside the perimeter of the ground used by aircraft, leading to a railway station or other surface terminal.

In fact, Morris Jackaman's concept was built at Gatwick to a detailed design by architect Frank Hoar, complete with the telescopic walkways referred to in the patent application and adjacent to and linked by tunnel to the Southern Railway station built adjacent. The first service operated on Sunday 17 May 1936 to Paris: passengers caught the 12.28 train from Victoria, arriving at Gatwick Airport station at 13.10. They mounted the stairs of the footbridge, crossed to the up platform, walked through the short foot tunnel, and completed passport and other formalities in the terminal ready for the 13.30 departure of British Airways' DH86. They left the terminal through

▼ 2.1 Gatwick Airport Terminal, 1936.

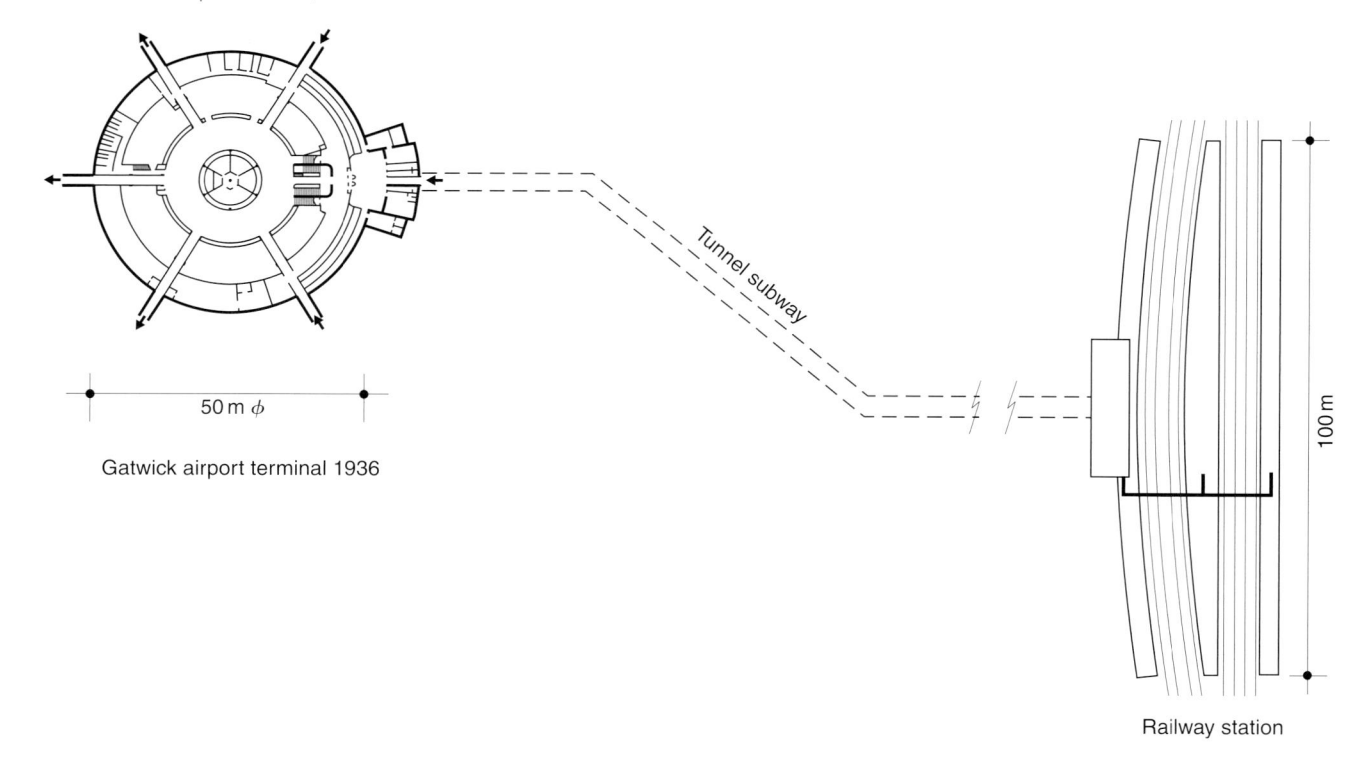

Gatwick airport terminal 1936
50 m φ

Tunnel subway

Railway station
100 m

2.2 Gatwick Airport, the flight boarding (courtesy of John King and Mrs Reeves (née Desoutter)).

2.3 Gatwick Airport, showing the railway station.

the telescopic canvas-covered passageway to board the aircraft steps. Ninety-five minutes later they reached Paris. The whole journey from Victoria had taken two-and-a-half hours and cost them £4-5s-0d, including first class rail travel from Victoria.

REFERENCES

Hoar, H. F. (1936). Procedure and planning for a municipal airport. *The Builder*, 17 April–8 May.

King, J. (1986). *Gatwick – The Evolution of an Airport*. Gatwick Airport Ltd/Sussex Industrial Archaeological Society.

King, J. and Tait, G. (1980). *Golden Gatwick, 50 Years of Aviation*. British Airports Authority/Royal Aeronautical Society.

2.2 Other successful multi-modal interchanges (see Chapter 6)

The four London airports of Heathrow (6.1.4, 6.1.5, 6.3.4), Gatwick (6.2.4), Stansted (6.2.3) and Luton (6.4.1) have all become multi-modal, with differing degrees of success. Others in Britain include Birmingham (6.3.2), Manchester (6.3.3) and Southampton (6.3.5).

The same pattern is repeated throughout the world, except that the greater the car usage, the less railway and bus usage and the fewer multi-modal interchanges.

3 The future development of integrated transport

Most cities have evolved separate land-based, air and, in some cases, waterborne transport over the years. The very separateness of these transport systems may have been the key to their viability in the face of logistical and physical constraints. Major cities like London and Paris have national rail and coach networks radiating from many stations, and both central and peripheral to their inner hearts. Despite the theoretical opportunities for riverborne transport linking other stations, the more universal solution of underground railway lines has been adopted. Only for waterside cities like Venice and Sydney can ferry systems make a real contribution. New low-density 'cities' like Milton Keynes in the UK in the 1960s and national capitals like Canberra and Abuja created in their planning networks of roads, bus-ways, etc., but high-density cities are not by definition planned from scratch.

3.1 Resolution of complexity of transport systems caused by dis-integration

First, we examine particular modes, why they demand inherent complexity, and compare air and rail.

3.1.1 Air

Why should an airport terminal be so complicated and ultimately so expensive? Why should a terminal be designed around the baggage system, and an ever more complicated one at that? Why should baggage handling be so? Because in the early days, air travel was for the rich few who could not take their flunkeys with them – so the airlines and airports doffed their caps and dealt with the cases. Conversely, rail travel, even transcontinental rail travel, was never like that. As far as is known, no airport or airline has attempted the railway system.

The reason lies with the aircraft. Boeing in the mid-1970s announced the 7 × 7, which became the 757 and 767, and one aim was to increase, or allow airlines to increase, the cabin baggage capacity, because that was thought to be a desirable option. The outcome was that 'bums on seats' were more important than eliminating a ground-level problem, and so the necessity for security-conscious ground-level baggage handling has escalated to the costly and high-technology activity which it is today.

Incidentally, in this area, one should look to the former USSR. Even in the 1990s one could see enormous, bursting cardboard suitcases being searched in cavernous, depressing terminals and wrapped in brown paper to prevent tampering before being lugged across snowfields by straggling lines of peasants. They loaded their own cases into the hold of the then 'new' IL-86 before climbing spiral stairs inside the fuselage to fly, without in-flight catering, for 14 hours across Soviet airspace to Vladivostok. How about that?

Considering security in more detail – the 'problem' has snowballed out of control! Having separated the airline passenger from his or her bag before it is effectively searched, systems have to be provided for reunion in the event of suspicion and for reconciliation of listed bags in the hold and passengers in the cabin before departure. This is far from ideal, and the only counter-measure can or will be imposed by the no-nonsense no-frills airlines at basic airports like Lubeck. There, all size- and weight-limited bags (and passengers) are X-rayed on entry to the check-in hall, so there is no clever outbound baggage system – the bags go on the truck behind the desk. On arrival, while the passengers queue on the tarmac at the door of a tiny immigration room, the bags go on a single flat carousel – opportunities for theft are minimal. Passengers wanting to take big items have to send them separately as cargo. British Railways in the 1950s and 1960s offered PLA (Passengers' Luggage in Advance), which was collected from home before departure and delivered at the other end.

Meanwhile, monster showpiece airport terminals are still being built and passengers are paying more in many cases for terminal facilities than for flights. Why should hand-baggage-only passengers pay for giant baggage handling systems, for example?

3.1.2 Rail

In the larger cities, radial networks emanate from the outer city ring because railway companies could get no closer to the heart to build their termini.

In a similar way, railway builders often could not penetrate town and city centres, due to lack of available 'corridors' for the wide curves needed, and built stations at the edge of the centre.

3.1.3 Converging standards

With the belated entry of low-cost airlines, standards of travel by air, rail and coach are converging, and with them the terminals which serve these three modes.

Illustrations are to be found throughout the adjacent remote airport interchanges in Chapter 6, with British airports like Birmingham, Luton and Southampton.

3.2 Problems in resolving dis-integration

3.2.1 Volatility

In Europe, for example, the study 'Future Development of Air Transport in the UK: South East', dated February 2003, does not sufficiently acknowledge the volatility of the market for air travel and its providers. The case is not made for more concentration of air travel at a few non-hub airports. Diversity and convenience is the key. Witness new dispersed airport activity all over Europe driven by no-frills carriers.

The major effect of these changes is only acknowledged by the stated view that any disincentive to fly imposed by new environmental and fiscal legislation will or may be offset by price cutting by no-frills carriers.

3.2.2 Access factors

Rail and coach stations go where the demand is, but rail operators have taken second place in towns and cities to buses and coaches, which can weave into the medieval and pre-Victorian urban fabric.

Airport locations have been determined by many historic factors, some of which have contributed to accessibility and others not. In the South East of England, for example, Luton Airport has always suffered from acute access problems, sufficient to limit its role. Heathrow has been progressively surrounded and constrained over the last 50 years. Stansted and Gatwick are inconveniently located, but provide valuable services, especially Stansted as the 'dumping ground' for no-frills point-to-point routes. As the quest for interchange intensifies, access becomes more critical. Heathrow has by far the greatest catchment area.

3.2.3 Transport operator factors

Rail operators are seeing the value of stations that serve the car user. Parkway stations outside Bristol and Southampton, for example, as well as serving the local area, also provide a park-and-ride facility and, in the case of Southampton, a fly-and-ride service. Stations around London on the principal radial rail lines at Stevenage, Watford and Reading are served by long-distance trains. The main TGV line north-to-south through France doubles at Lyon to provide a station at St Exupéry Airport.

Airlines have the greatest single impact on the airport market. The market is now polarised between the intercontinental carriers and the medium-haul no-frills carriers. Big birds are fed by small birds. A hub ceases to be a hub if the small birds cannot get in.

3.2.4 Planning procedures and responsibilities

While designed to protect the public and the environment, procedures are coming to be weighted in favour of the implementation and improvement of public transport.

The tendency in democratic society to veer away from totalitarian control of transport infrastructure to fragment responsibility is something that impedes progress, but a new cycle of private investment is starting to 'deliver'.

For example, in the UK, local roads are the responsibility of the local highway authority, motorways are the responsibility of the national Highways Agency, and rail, bus and air transport are principally private.

3.2.5 Conflicts of interest

Apart from conflicts that are resolved by public enquiries and planning procedures in general, there are examples of technical considerations outweighing an otherwise desirable development.

Motorway traffic management militates against park-and-ride sites being placed at motorway intersections.

Space and resources are inevitably limited and the role of the transport planner is to reach the best compromise.

3.3 Examples

3.3.1 European rail–air integration: the train that thinks it's a plane

Rail–air substitution is a reality in Europe, either for airport-to-airport journeys or for city-centre-to-city-centre journeys. High-speed trains in Europe are linking airports direct to cities which would otherwise have been connected by air journeys and train connections to city centre.

Forty per cent of journeys in Europe are less than 500 km, so can be made by train, reducing pressure on airports. Note that the advent of high-speed rail from London to the Channel Tunnel, and thence to Paris, Amsterdam and Brussels, is predicted to take significant traffic from airlines (see Chapter 6.6.5).

Other current examples include:

- Paris–Lyon St Exupéry Airport;
- Cologne–Frankfurt Airport – Lufthansa leases seats on trains;
- Brussels–Paris.

3.3.2 An example of the problems in national strategy: a single national hub airport for London

Neither passengers nor airlines will benefit from splitting 'hub' demand between airports. Intercontinental carriers or global alliances of airlines need to operate as economically as possible, so they will not easily provide services from more than one international airport in the South East of England. It is not fanciful to specify that London should have a single hub with a capacity of over 150 million passengers per annum. The answer lies at Heathrow. Or, failing that, with land links between Heathrow and other airports to feed it, and vice versa. Coaches racing round the M25 are not the answer. A proposed single short runway to the north of the existing Heathrow site is not enough, so what Heathrow needs is a satellite airport. This concept needs to be combined with the runway capacity solution. The optimum site, therefore, is absolutely not in the Thames Estuary, nor even as remote as Gatwick, Luton or Stansted, which are all well over 60 minutes by road from Heathrow. This presumes, reasonably, that augmented land links between these sites are not

3.1 London Oxford (LOX) satellite airport location, showing other airport sites and proposals in the UK South East Region (Courtesy of Pleiade Associates – see website www.pleiade.org).

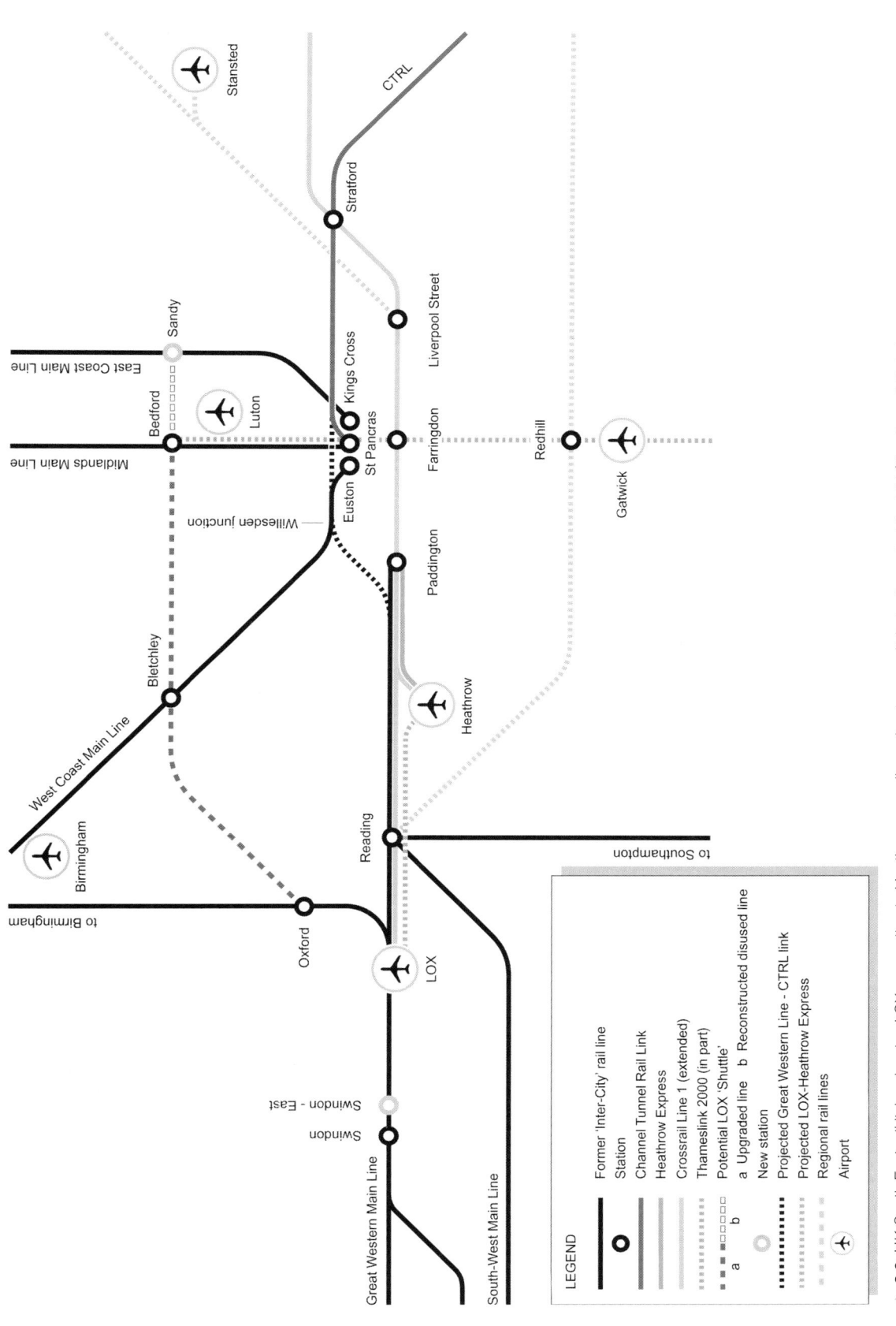

Transport terminals and modal interchanges

3.2 UK South East rail links, showing LOX connection to Heathrow as well as advantageous links to the rest of the UK (Courtesy of Pleiade Associates – see website www.pleiade.org).

The future development of integrated transport **11**

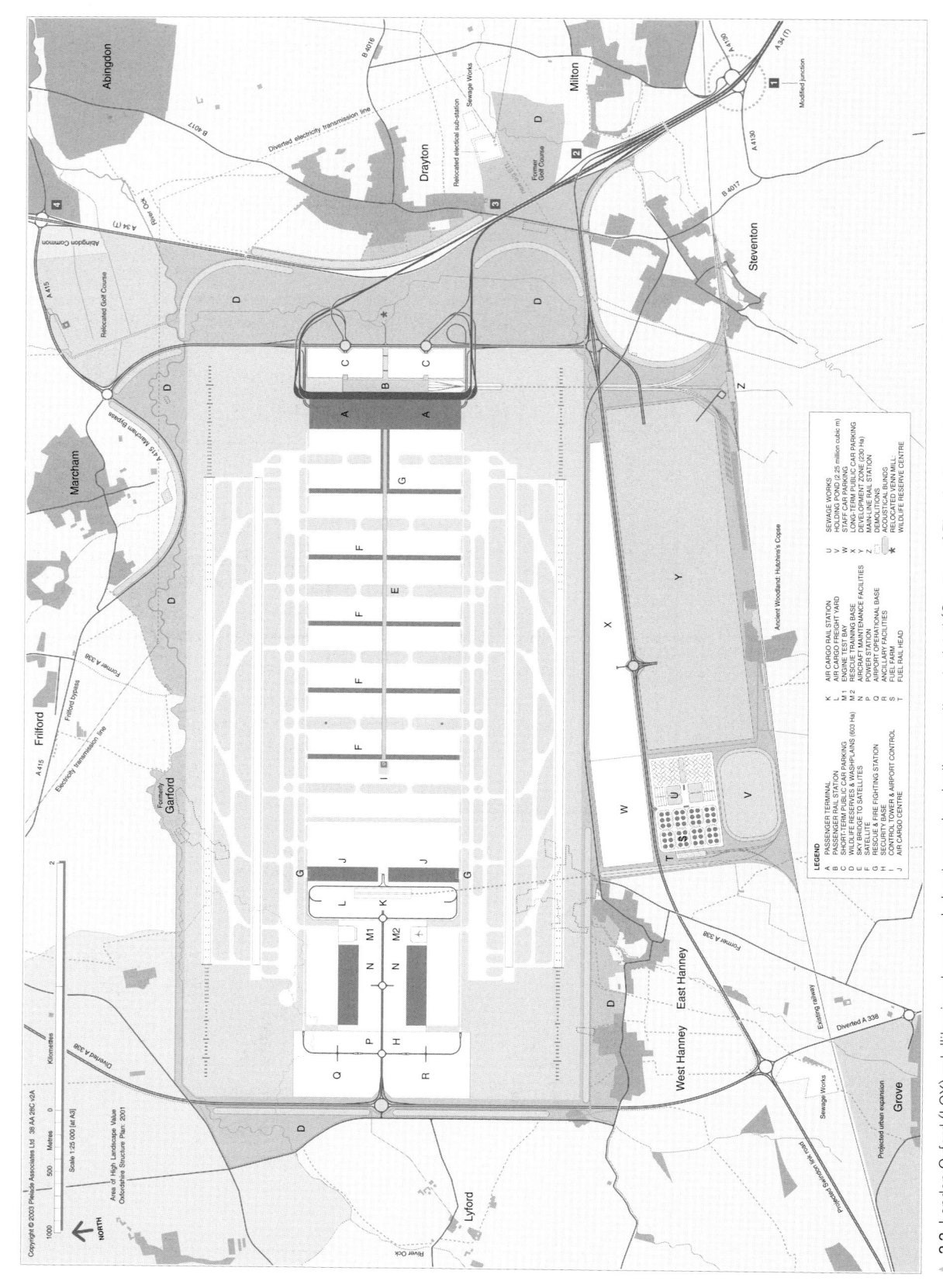

Transport terminals and modal interchanges

3.3 London Oxford (LOX) satellite airport proposal, showing road and rail connections in detail (Courtesy of Pleiade Associates – see website www.pleiade.org).

12   The future development of integrated transport

possible for the high-quality, quick-flight transfers achievable at Frankfurt, Amsterdam or Paris. For economy of travel distance and time to the maximum British population, the satellite should be located north and west of Heathrow. It should be, say, 20 minutes from Heathrow, significantly closer in travel time if not necessarily in distance than Gatwick, Luton and Stansted.

London Oxford

This fully-developed proposal was presented in 2003 – the site having been chosen for its advantages of minimum disruption and displacement and maximum transport infrastructure connection, including air traffic corridors.

Such a development would circumvent problems of disruption and displacement, traffic congestion, pollution and planning delays at Heathrow.

3.3.3 Joining a national rail network comprehensively to an airport: Grand Junction Link

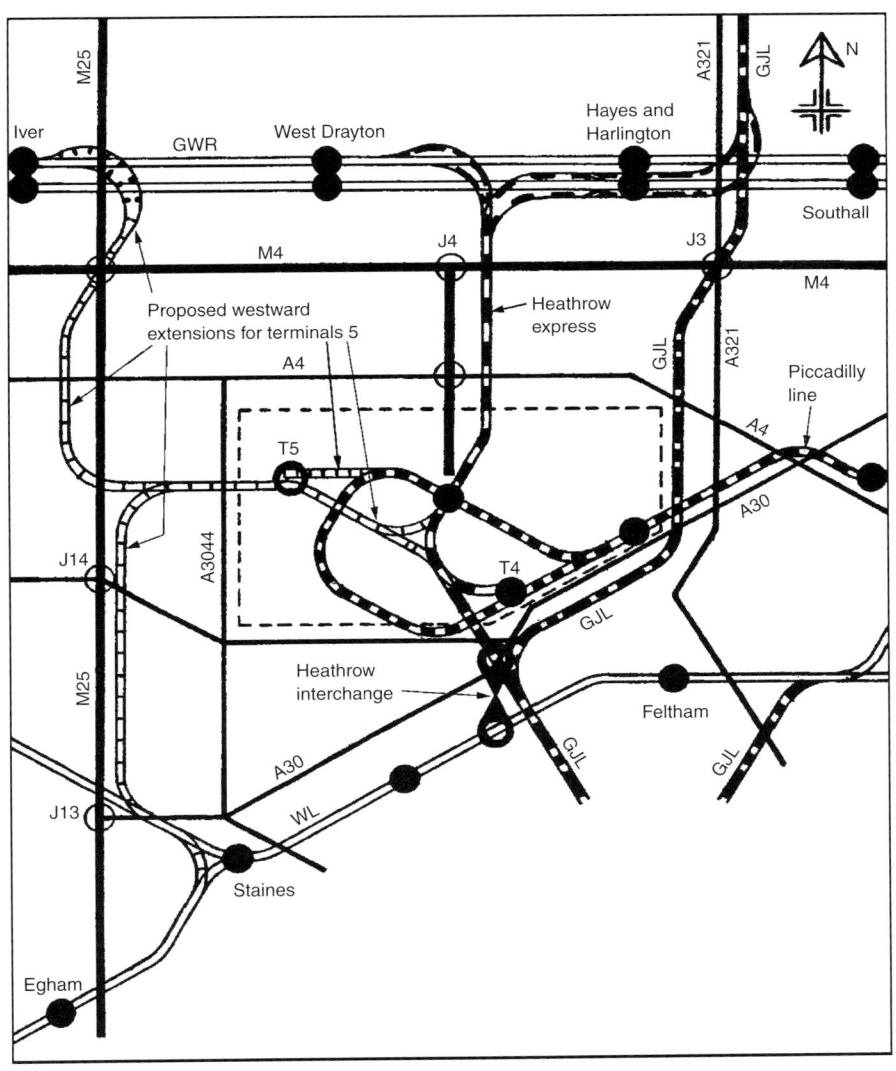

▲ 3.4 Grand Junction Link and other railways around Heathrow Airport.

Transport terminals and modal interchanges

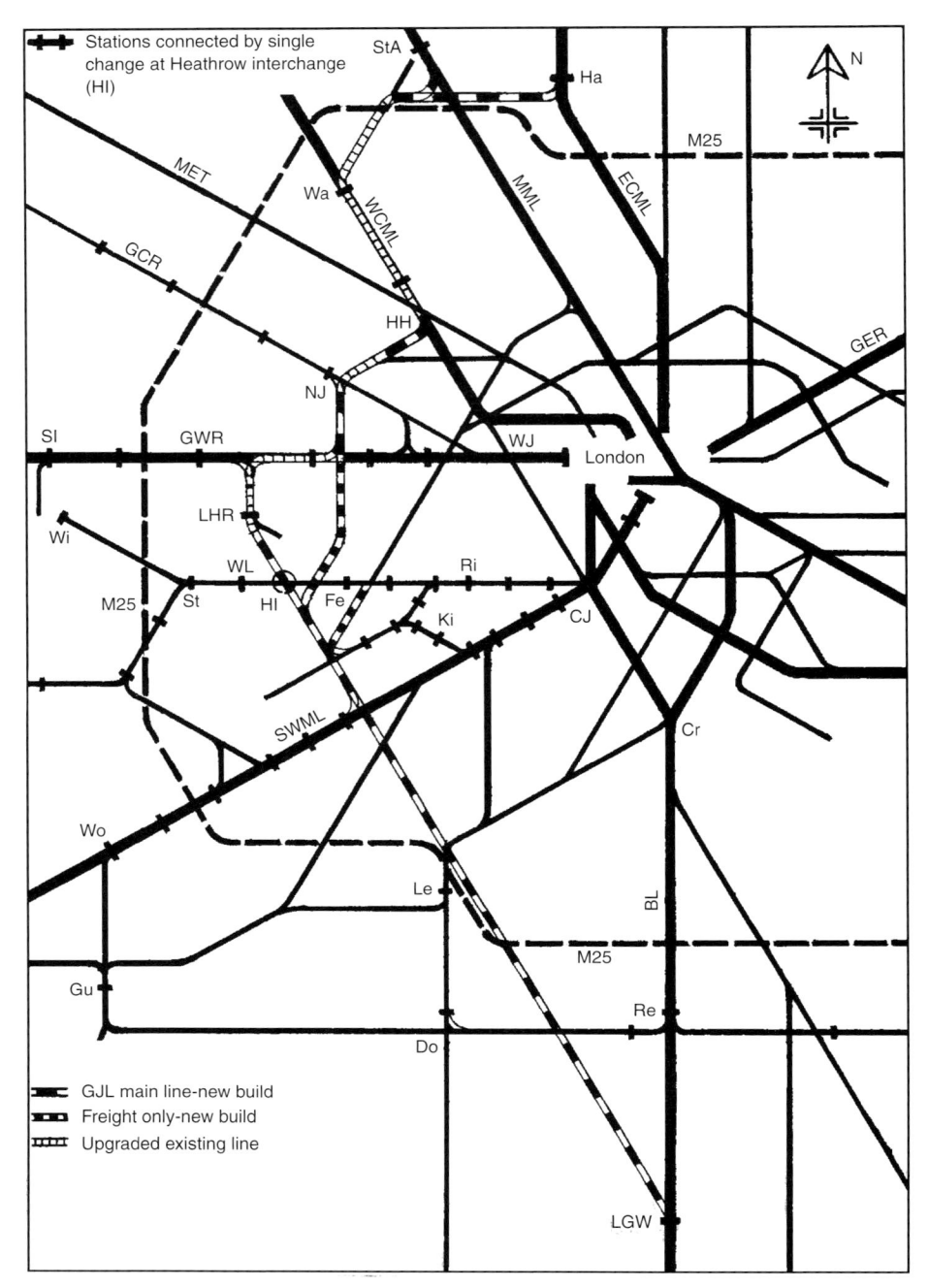

▲ 3.5 Grand Junction Link and existing networks: Heathrow connections. This shows the ten radial railways from the ten London termini inherited from the days of Victorian railway building and the effect of the GJL in creating an orbital connection of eight of the ten.

Heathrow Airport, starting in the late 1940s beside the Bath Road heading west and 16 miles distant from Central London, was long unserved by rail services in spite of being a stone's throw south of the Great Western main railway line linking London and Bristol. First, the Piccadilly Line of the London Underground and then a dedicated heavy rail spur to London Paddington have offered competition to road vehicle access. Comprehensive access to the

14 The future development of integrated transport

national railway network is far off, but a far-sighted proposal has been developed to link main lines south, west and north of London and Heathrow. While not achieving the truly multi-modal hub, which Central London can never be, the addition of a line called the Grand Junction Link (GJL) comes nearest to putting Heathrow on the national rail map and vice versa.

REFERENCES

Aviation Environment Federation (2000). From planes to trains: realising the potential from shifting short-haul flights to rail, Report for the Friends of the Earth.

Booz-Allen & Hamilton (2000). Regions to London and London's Airports Study, Report for the Strategic Rail Authority.

Elliff, C. (2001). Rails around London – in search of the railway M25. *Transport 147 Proceedings of the Institution of Civil Engineers*, London, May.

Website: www.pleiade.org

4 Two particular studies point the way

4.1 UK Highways Agency study, 'Underground Transport Interchanges'

Although this study, 'Underground Transport Interchanges', focuses on some possibilities to conceal facilities underground, it develops a series of models independent of whether they are at or below ground level. These models include road-to-rail interchange, coach hub and rail station forecourt, and are reinforced by eighteen best practice criteria.

CREDITS

CLIENT: The Highways Agency
TEAM: Halliburton KBR, YRM Architects Designers Planners, Hewdon Consulting

4.1.1 The rail parkway

A rail-based park-and-ride interchange mainly serving long-distance travel. The interchange could include underground parking, spacious circulation and provision for a light rail terminus.

4.1.2 The coach hub

This model envisages a network of high-speed, high-frequency coach services running on motorways or trunk roads, possibly in dedicated bus lanes, picking up passengers at airport-style terminals at special interchanges.

4.1.3 The rail station forecourt

This model provides hire car service (using environment-friendly vehicles) at the rail station, with the aim of reducing long-distance car trips. Cars would be stored in silos, which could also be used for light freight containers received from trains for distribution in town centres by road.

4.1.4 Eighteen best practice criteria, to be applied at early design stage

1 An interchange should be an uplifting experience.
2 The design should emphasise the interchange's role as a portal into different transport modes, provide a welcoming environment for the traveller and create interest by emphasis of the arrival and departure points.
3 Natural daylight should be maximised wherever possible, creating a sense of well-being and reducing a sense of enclosure.

▲ 4.1 The rail parkway: circulation area and LRT platforms (Figures 4.1–4.6 courtesy of Halliburton KBR/YRM Architects Designers Planners).

▲ 4.2 The rail parkway: vertical integration of transport modes.

Transport terminals and modal interchanges

▲ 4.3 The coach hub: open, spacious and secure environment.

4.4 The coach hub: coach concourse.

▲ 4.5 The rail station forecourt: transfer of freight to underground silo.

▲ 4.6 The rail station forecourt: station kerbside.

4 Imaginative use of lighting can give many opportunities for holding interest, and variety and colour enhance this effect.
5 Consistently high levels of passenger comfort should be provided both in the interchange and in the transport itself.
6 Good air quality is important, however enclosed the space, above and below ground.
7 Spaciousness is important, especially since many people are prejudiced against enclosed and underground stations because these spaces have been cramped in the past.
8 An interchange should be designed with good sight lines between different modes of transport, to assist wayfinding and add a sense of interchange experience.
9 Soft internal landscaping will soften the design of interiors and provide variety.
10 The geometry of the interior and choice of materials should ensure a calm interior.
11 The architecture, technology and facilities should work together to provide a coherent whole.
12 The architectural expression of the interchange should reflect the culture of this century and the technology of contemporary travel.
13 Fun and drama should be injected into the experience.
14 Sculptures and fountains should be introduced to act as focal points.
15 Each interchange should have a distinct identity, though with consistency in the design of elements such as wayfinding to make movement easy to understand.
16 High-quality construction should ensure that interchanges are desirable places to visit for a long time.
17 Design should be 'timeless' yet of its time.
18 Robust design should give the interchange a sense of permanence.

REFERENCES

Highways Agency (2002). Underground Transport Interchanges, unpublished.
Scholey, J. (2003). Intermodal underground. *Passenger Terminal World Annual Technology Showcase Issue*.

4.2 The Scott Brownrigg/RCA InterchangeAble research programme: reclaiming the interchange

Urban multi-modal interchanges can only benefit the community by making public transport more attractive and therefore more viable at the expense of private transport, as well as opening up commercial and social opportunities by greater throughput. The 'station' can be a more vital and synergetic public facility, as is demonstrated by the study.

CREDITS

CLIENT: Scott Brownrigg and the Helen Hamlyn Research Centre at the Royal College of Art, London
TEAM: Fiona Scott, Research Associate

4.2.1 Background

Scott Brownrigg, as architects, urban designers and planners, are proactive in showing the feasible physical solutions, as for hubs anywhere, be they primarily railway stations, airports or motorway stations. The very attractiveness of these solutions will prove complementary to the British government's aim to streamline, rather than curtail, the consultation process involved in major infrastructure projects.

Early investment in the development of design ideas particularly supportive to the replacement of corrosive private transport by sustainable public transport will pay dividends. Both special interest groups and the public at large will see the benefits more clearly when commenting upon or even criticising 'little local difficulties'.

All have a part to play, transport operators, local and central authorities, developers, investors and designers, but it is the vision of designers which will act as a catalyst to both the process and the product.

The process:

- Proactive design.
- Demonstrable alternatives.
- Social benefits.

The product:

- Improved public transport, accessible to all and attractive to all.
- Reclamation of the 'station' as a thriving public space.
- Retention/enhancement of nineteenth-century structures which founded the British railway system (see section 6.6.5 in Chapter 6).
- Quality urban design.

4.2.2 The study

Key mobility functions were analysed, and the interchange placed at the focal point of a series of factors.

Six specific issues and solutions were identified, with user scenarios to back them up:

1. Disconnection and the urban connector (see Figures 4.8 and 4.9).
2. Inaccessibility and urban ribbons (see Figures 4.10 and 4.11).
3. Public spaces and the community hub (see Figures 4.12 and 4.13).
4. Low density and the stacked programme (see Figures 4.14 and 4.15).
5. In-between spaces and residual space-makers (see Figures 4.16 and 4.17).
6. Information (see Figures 4.18 and 4.19).

The following twelve illustrations are paired – a telling image of each problem issue is followed by a pictorial representation of a strategy to reclaim the interchange for quality public space, facilities and transport enhancement.

REFERENCE

Scott, F. (2003). *InterchangeAble*. Helen Hamlyn Research Centre, Royal College of Art, London.

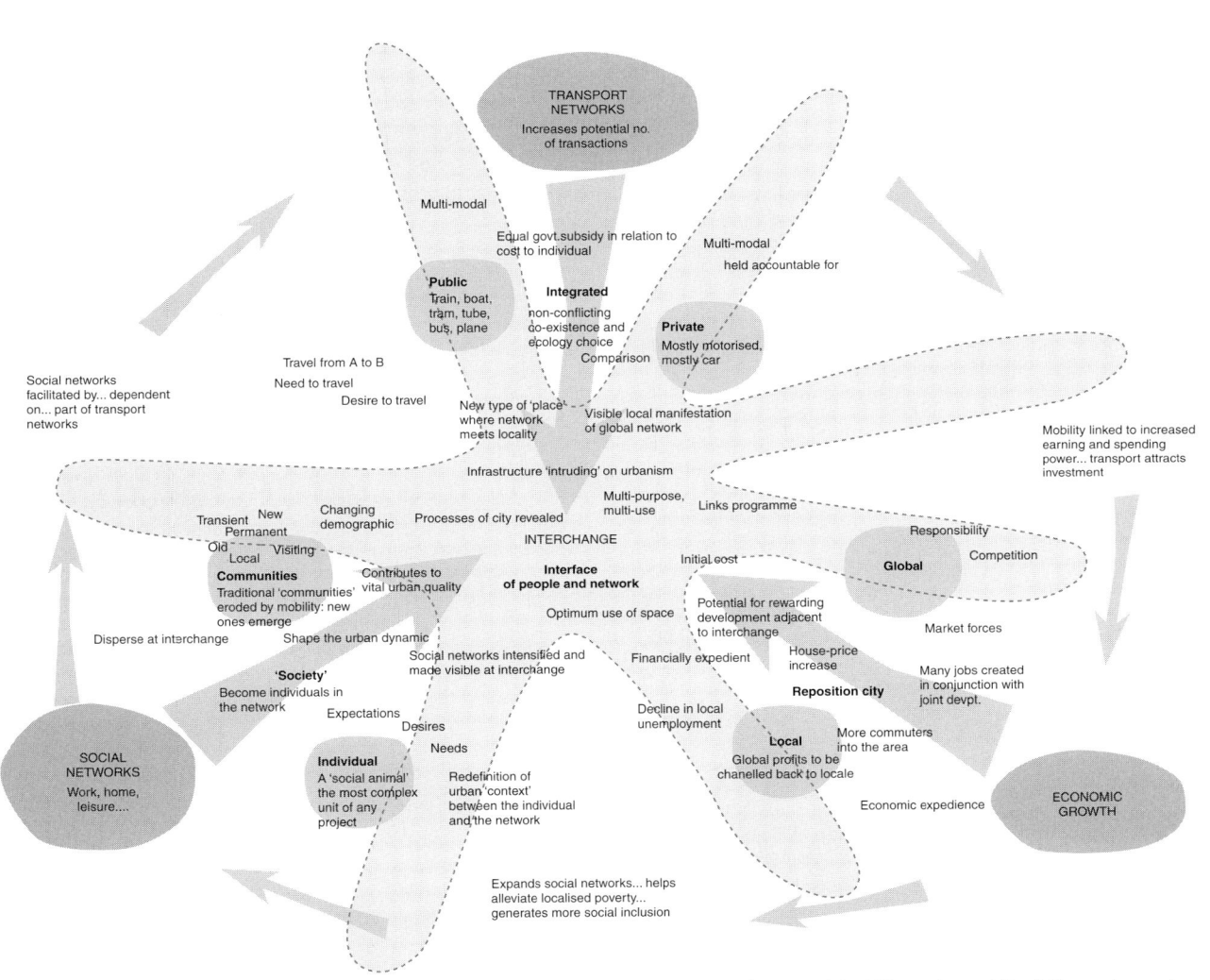

Transport terminals and modal interchanges

4.7 Key mobility functions (Figures 4.7–4.19 courtesy of Scott Brownrigg and the Helen Hamlyn Research Centre at the Royal College of Art).

Two particular studies point the way   21

Transport terminals and modal interchanges

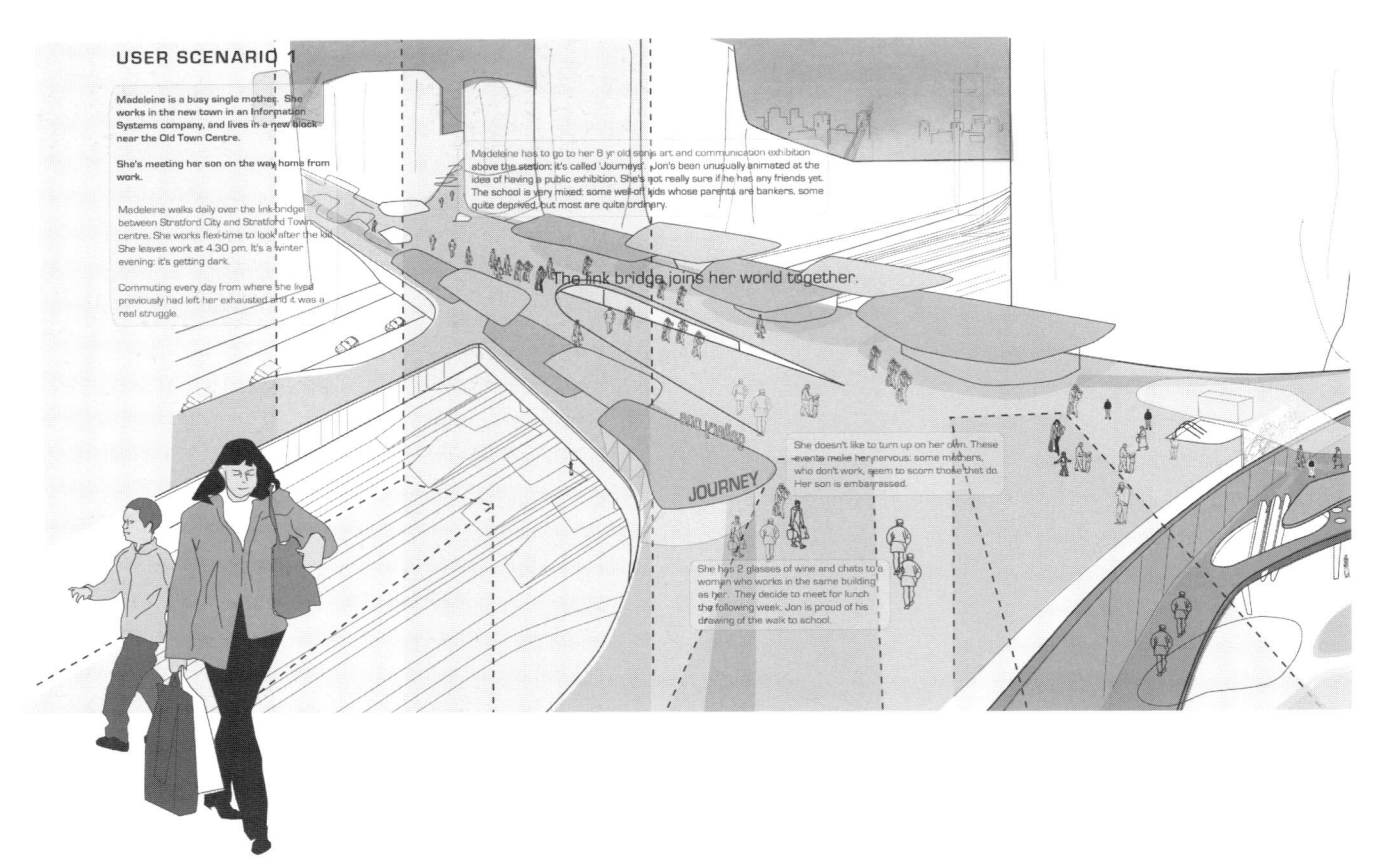

4.8 The urban connector. A pictorial user scenario: a strategy for mending the breaks in urban fabric caused by rail lines and roads. The connector can host retail and other facilities to promote pedestrian journeys and create a local landmark.

4.9 Disconnection and the urban connector. Transport corridors can be barriers.

Two particular studies point the way

Transport terminals and modal interchanges

▲ 4.8 (Continued)

▲ 4.10 Inaccessibility and urban ribbons. Public transport can be inaccessible to the pedestrian.

Two particular studies point the way 23

Transport terminals and modal interchanges

4.11 Urban ribbons. A pictorial user scenario.

24 Two particular studies point the way

Transport terminals and modal interchanges

▲ 4.12 Public spaces and the community hub. Desolate public space. A missed opportunity for transport users.

USER SCENARIO 3

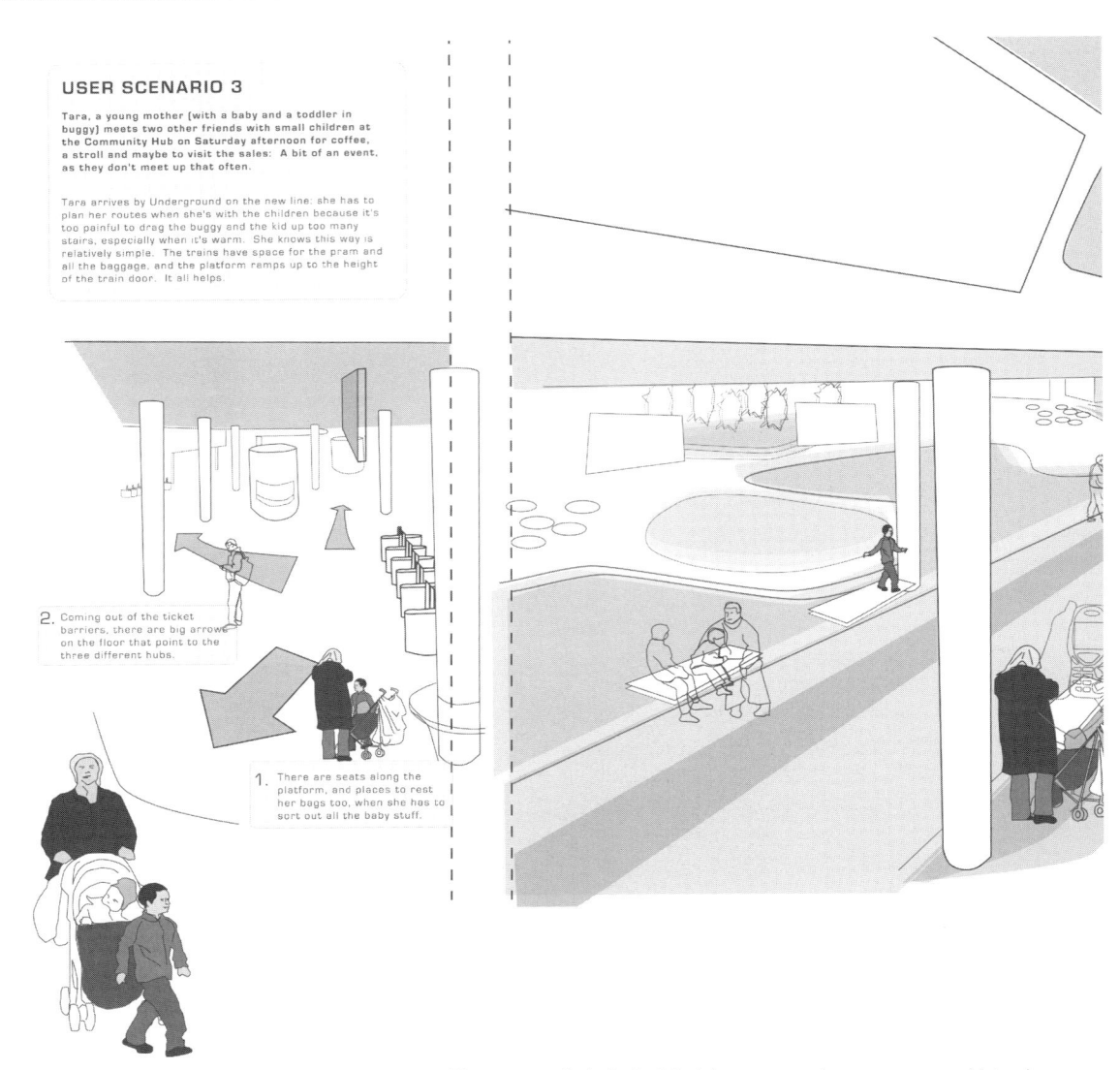

Tara, a young mother (with a baby and a toddler in buggy) meets two other friends with small children at the Community Hub on Saturday afternoon for coffee, a stroll and maybe to visit the sales. A bit of an event, as they don't meet up that often.

Tara arrives by Underground on the new line; she has to plan her routes when she's with the children because it's too painful to drag the buggy and the kid up too many stairs, especially when it's warm. She knows this way is relatively simple. The trains have space for the pram and all the baggage, and the platform ramps up to the height of the train door. It all helps.

2. Coming out of the ticket barriers, there are big arrows on the floor that point to the three different hubs.

1. There are seats along the platform, and places to rest her bags too, when she has to sort out all the baby stuff.

▲ 4.13 The community hub. A pictorial user scenario: spaces around interchanges do not need to be uninviting.

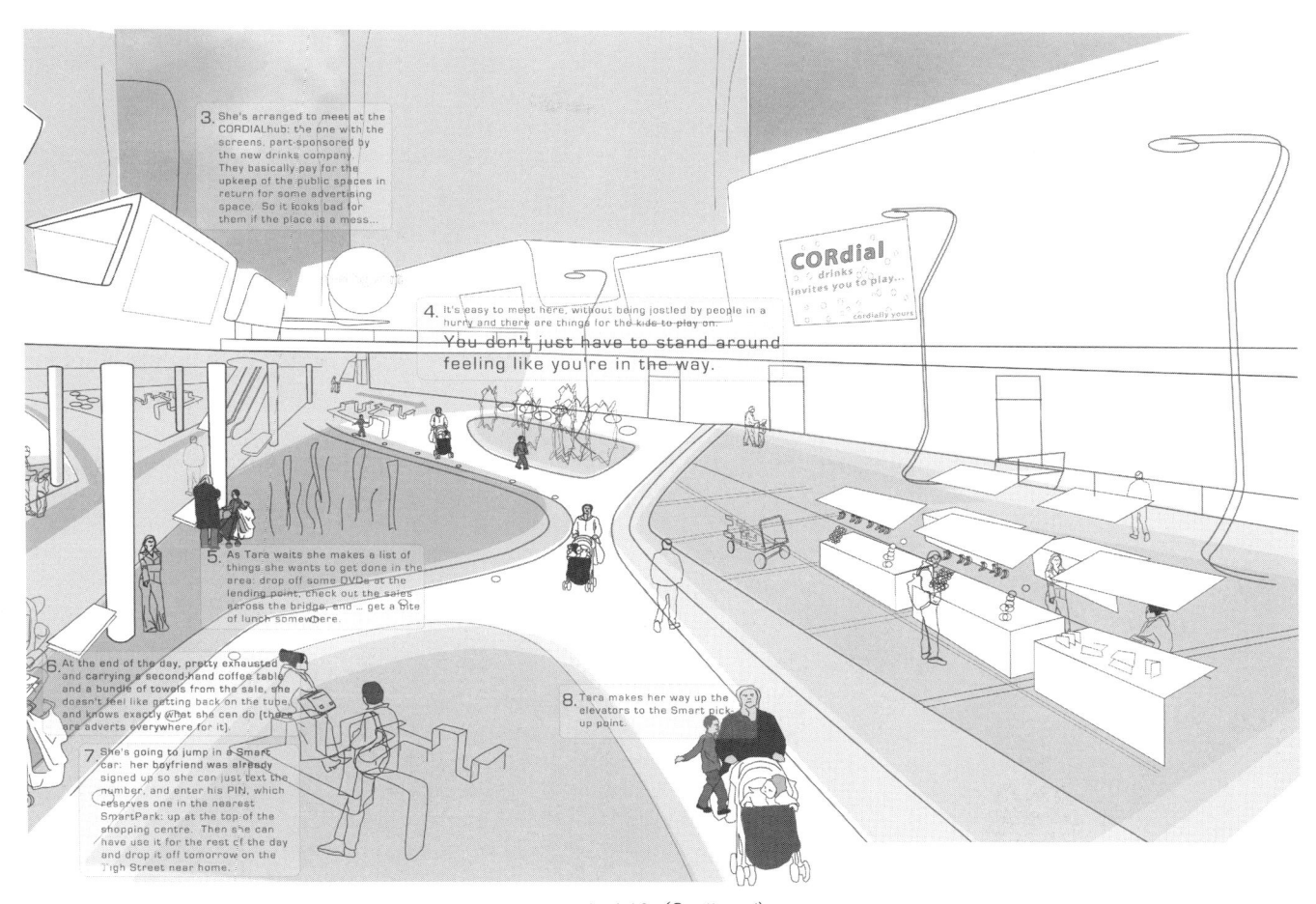

▲ 4.13 *(Continued)*

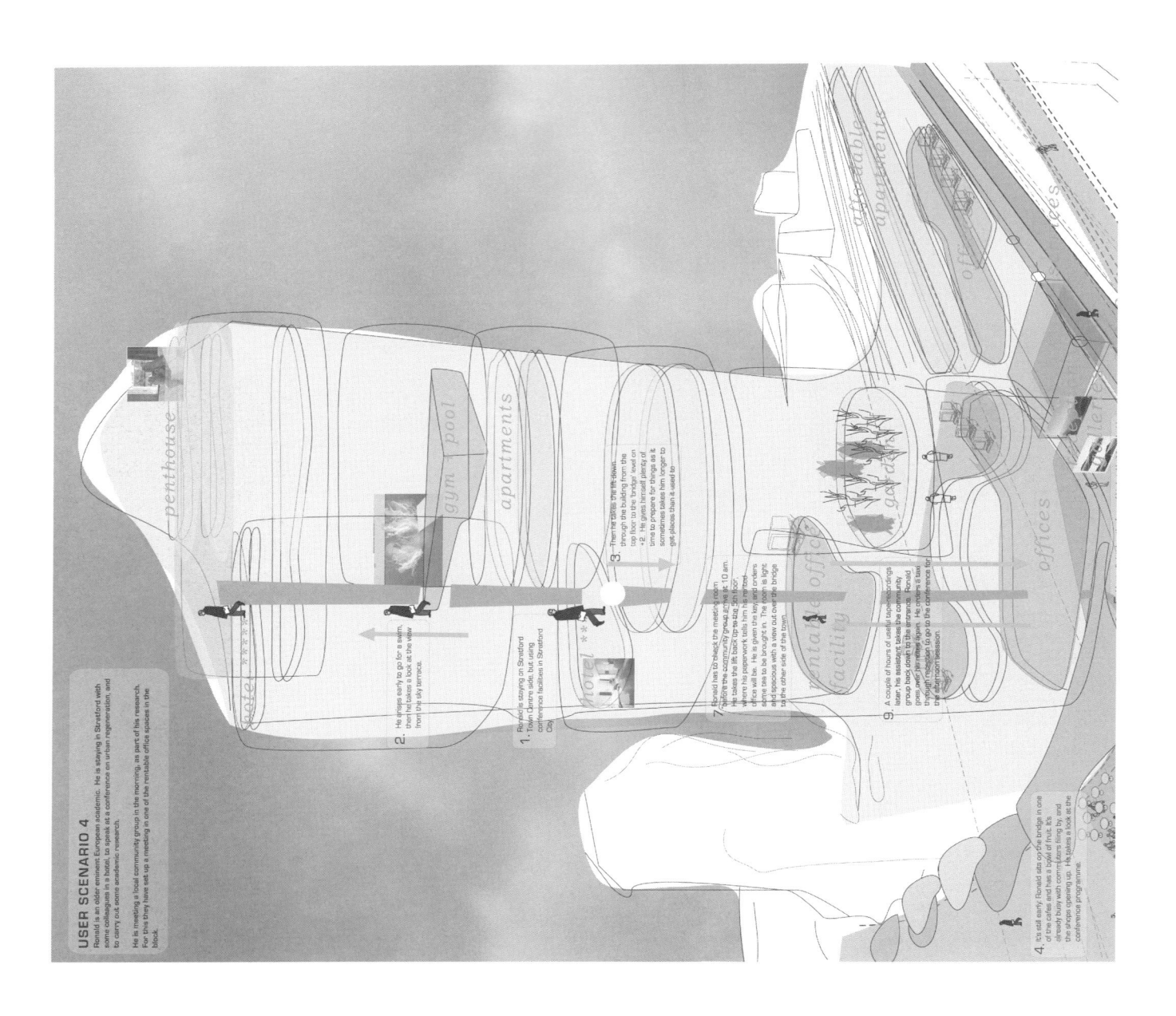

Transport terminals and modal interchanges

Transport terminals and modal interchanges

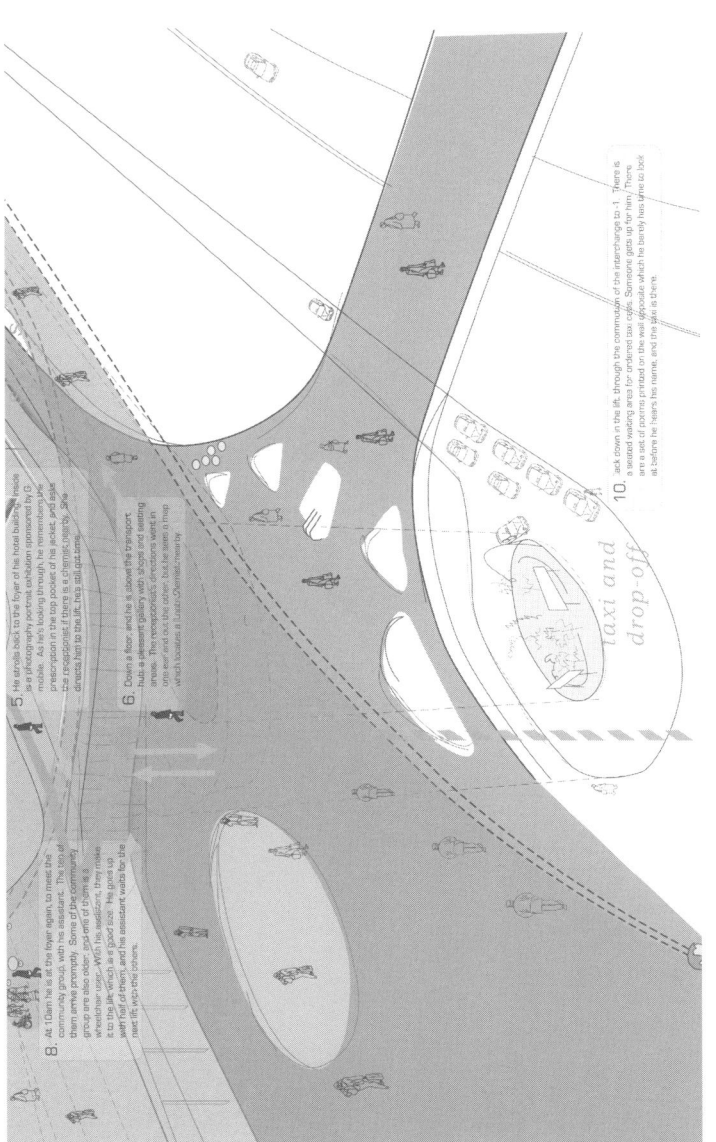

4.14 A pictorial user scenario: dense mixed use, creating interfaces with adjoining buildings in both horizontal and vertical planes.

4.15 Low density and the stacked programme. Wasted commercial opportunities.

Two particular studies point the way **29**

USER SCENARIO 5

Laura has her own design company. She rents in the live-work Village set-up that recently opened behind the interchange. To encourage small local businesses, and also to cut down on travel, the deal offers shared work-space on the ground floor, and very affordable apartments above. There are also workshop-type units.

This evening, Laura is going to a fencing lesson at the therapeutic sports centre on the other side of the station.

On the other side is a lift up to the podium level. If this isn't working there are at least two others further into the interchange that she can use.

It's a very sociable working environment here for a sole-practitioner, but the place is very adaptable for business growth until your business gets too big.

Laura finds the Village suits her well: she is a wheelchair user, and had previously found it hard to travel to her studio on public transport. The other facilities in the area are an immense bonus.

She goes through the art-tunnel under the embankment: this was designed through a competition won by a group of artists and designers from the Village, so they tend to look after it well.

It's also really easy for clients and manufacturers to visit her, and it looks good on her website to be so accessible.

4.16 A pictorial user scenario: even the most apparently unusable spaces can, with access, be of great value to the community.

4.17 In-between spaces and residual space-makers. Junctions typically result in pockets of under-used space.

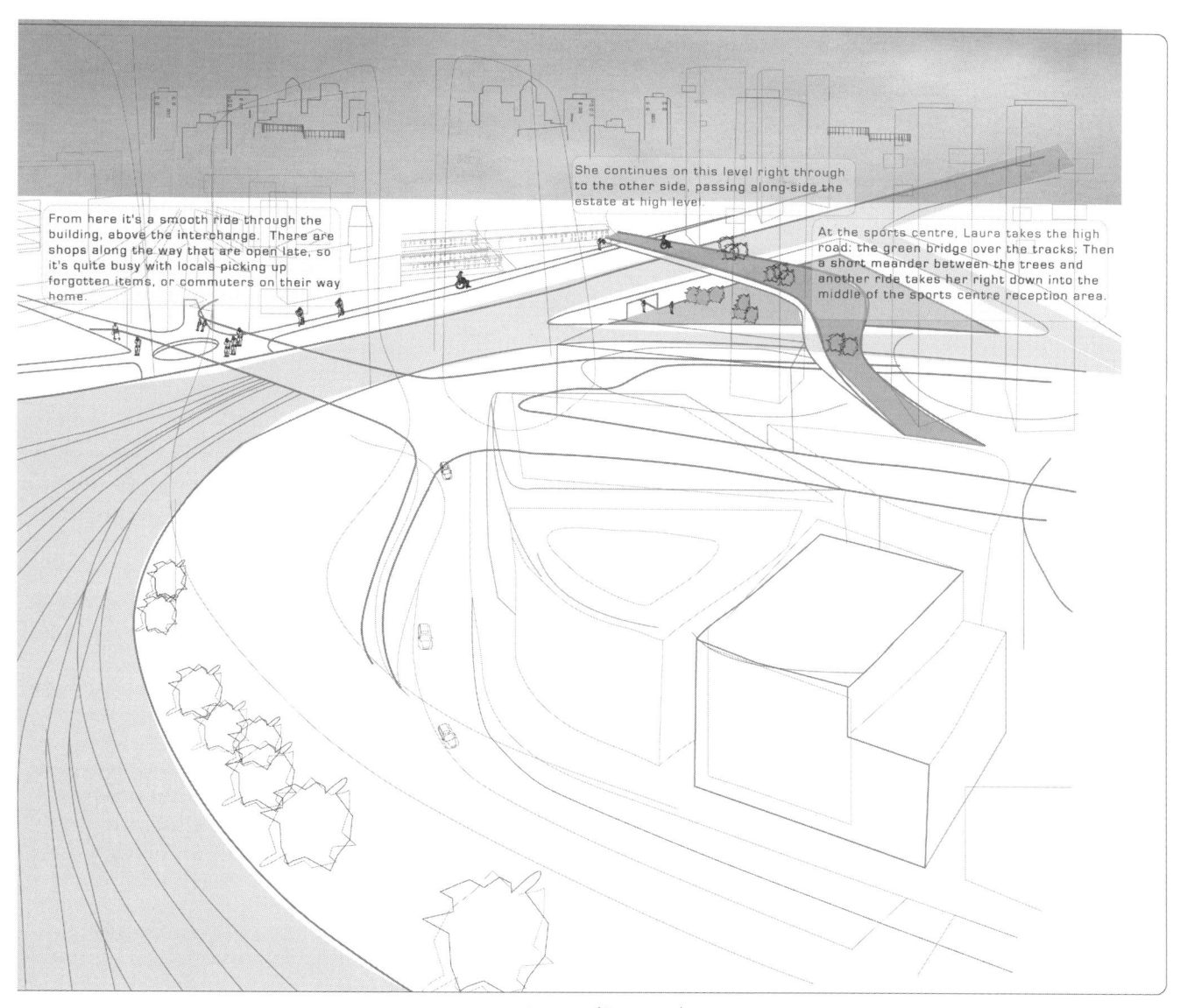

From here it's a smooth ride through the building, above the interchange. There are shops along the way that are open late, so it's quite busy with locals picking up forgotten items, or commuters on their way home.

She continues on this level right through to the other side, passing along-side the estate at high level.

At the sports centre, Laura takes the high road: the green bridge over the tracks; Then a short meander between the trees and another ride takes her right down into the middle of the sports centre reception area.

▲ 4.16 *(Continued)*

▲ 4.18 Information. Lack of co-ordination, resulting in clutter and confusion.

Transport terminals and modal interchanges

USER SCENARIO 6

Bill was expecting to take the underground straight to the club meeting, but heard on the radio that the local line was not working. This was confirmed when he was passed the bus-stop on the way to the corner shop, where the latest system news was digitally displayed.

He had consulted a map, but was still not clear what would be the best way to get into town. When he checked on the internet, it suggested cycling to the Edgetown interchange, and getting on the underground there. The website linked him through to a Route-Navigator site, which showed the most cycle-friendly way to get there from his street.

Half an hour later, he was approaching the interchange amongst a stream of other cyclists. The cycle-lane split off at some point, and descended beneath the station into the CycleCentre, which had been much publicised but which he had never seen.

Here, Bill found cycle shops and repair and servicing centres. There was a choice of cycle racks and lockers: he chose the option that was caged and manned, given that he could buy the travel-card there and then, and the lock-up was included in the price.

Unburdened of bike gear and helmet, Bill was a little disorientated, as he'd never been here, and it was the middle of the rush hour.

A coloured band on the floor and w only route up into the main interch alcove at the bottom of the stairs s to go, and told him to what to look

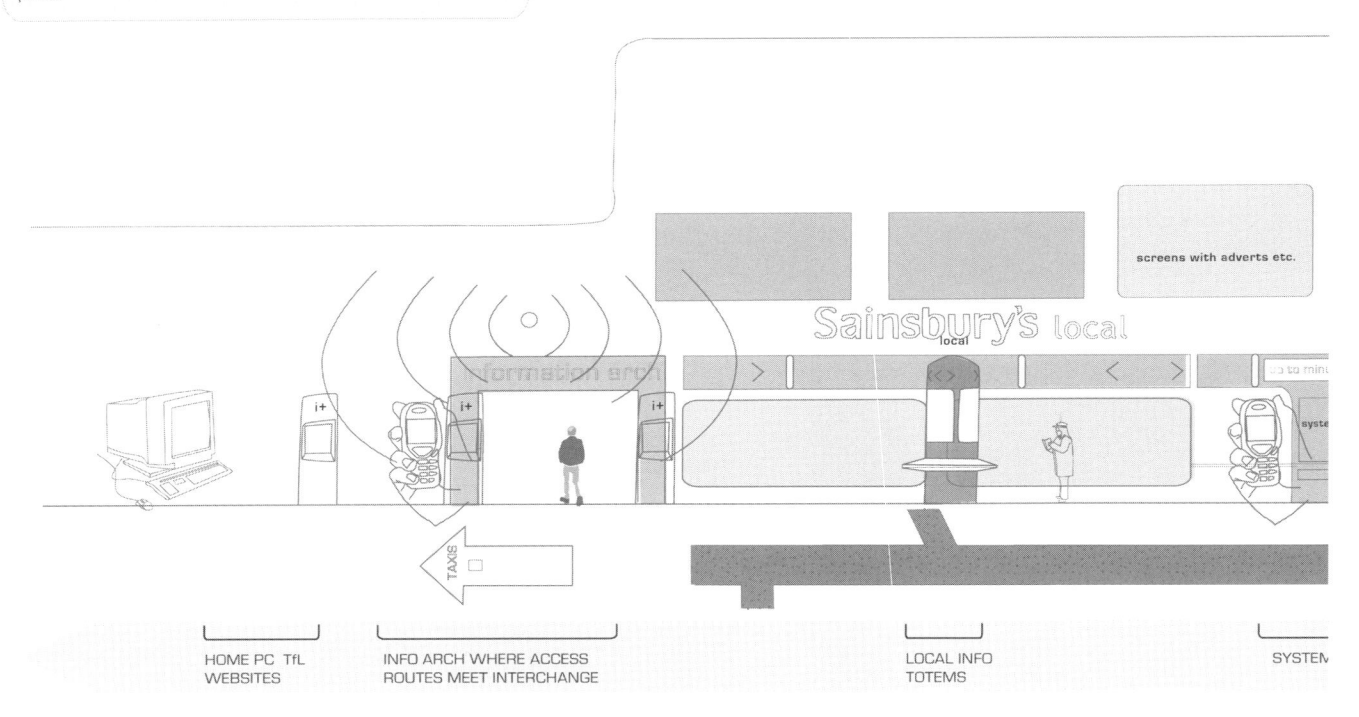

HOME PC TfL WEBSITES

INFO ARCH WHERE ACCESS ROUTES MEET INTERCHANGE

LOCAL INFO TOTEMS

SYSTEM

Sectional Diagram

▲ 4.19 A pictorial user scenario for info-flow: a hierarchical strategy for communications elements, to prevent piecemeal accumulation of layers of conflicting signals.

Two particular studies point the way

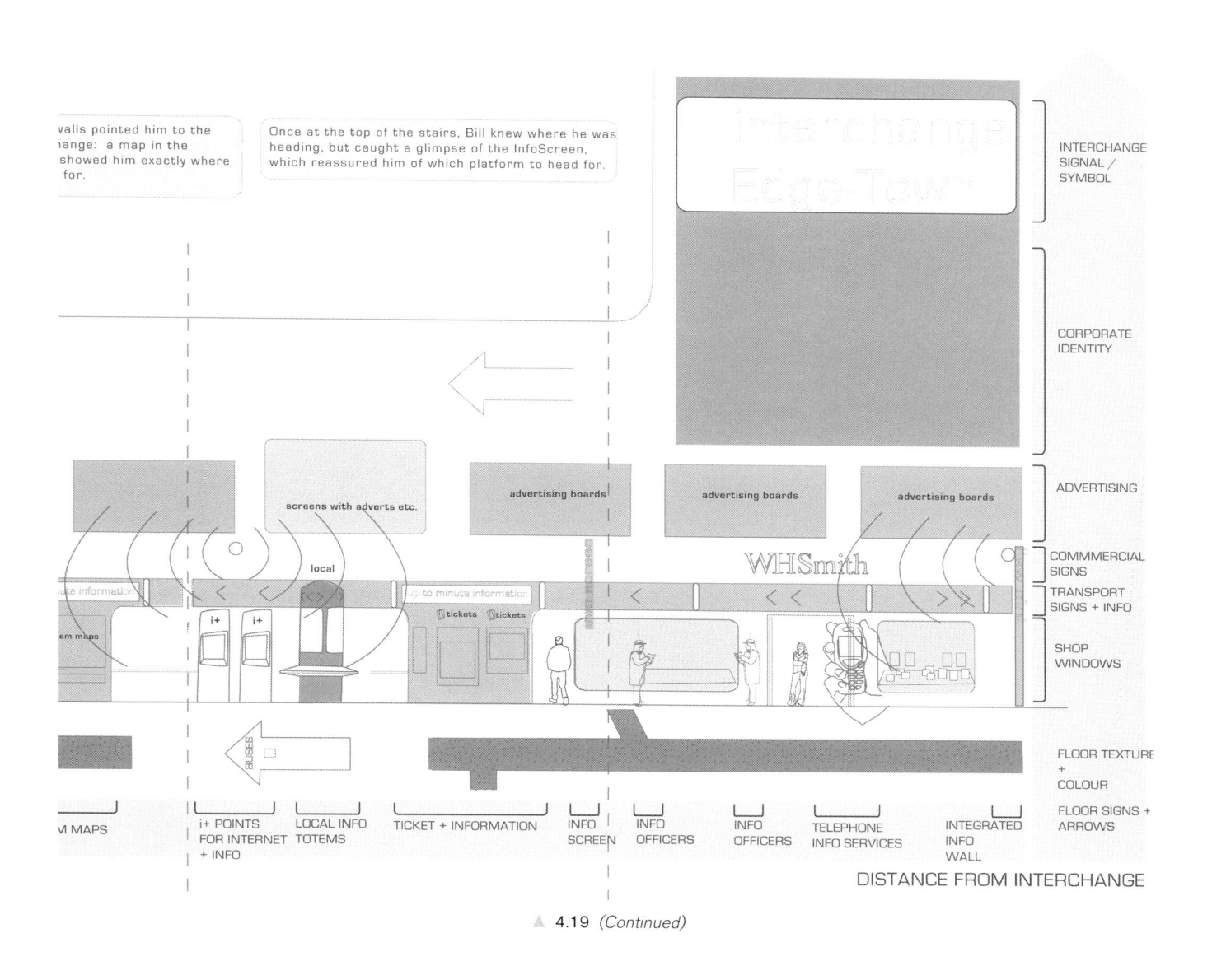

Transport terminals and modal interchanges

▲ 4.19 *(Continued)*

Two particular studies point the way **33**

5 Twenty-first-century initiatives

The pressure is on to make public transport attractive, with the long-term effect of improving quality of life in towns and cities (if not also the countryside and hinterland) and saving depleting energy resources. Considerable research energy has been applied to offer recommended ways of achieving successful transport interchanges. This is because it is not only the public transport vehicle and its service which counts: It is the whole journey, and that so often involves a traveller transferring from one route or mode to another.

It is worth reflecting on a pattern of transport usage in the UK, for example in the second half of the twentieth century (quoted from Solent Transport):

- Bus passenger kilometres per year have declined from 60 billion to 30 billion
- Rail passenger kilometres per year have been steady at 30 billion, whereas
- Car passenger kilometres per year have risen from almost nil to 600 billion.

Any step to make it possible to reduce that 600 billion kilometres per year is worthwhile.

Initiatives which focus upon the importance of the interchange concentrate upon processes such as consultation, collaboration and information rather than the built form and specific technologies. Some process-oriented recommendations are thwarted by the fragmentation of responsibility and the public/private debate referred to in Chapter 3.

In addition to the two studies quoted in Chapter 4, which focus upon design, the following initiatives are current and of particular interest.

5.1 International Air Rail Organisation (IARO)

This organisation exists to promote rail links with airports and has assembled an Internet database of 90 (as at Summer 2004) airports with rail links, 19 of them in the USA. This is to be found at www.airportrailwaysoftheworld.com.

REFERENCES

International Air Rail Organisation (2003). Yearbook.
Sharp, A. (2004). Developments in airport rail access. *Pan European Airports*, January.
Websites: www.iaro.com and www.airportrailwaysoftheworld.com.

5.2 International Air Transport Association (IATA)

The IATA *Airport Development Reference Manual*, 9th edition, gives the following lead on behalf of its members, for the first time embodying recommendations in its manual.

5.2.1 Recommendations, rail

Sound business and environmental case: the investment needed to provide dedicated airport rail provision can be very substantial. The business case should consider:

- Cost to the airport to provide the rail system.
- Cost to the airport not to provide the rail system.
- Public perception of the usefulness of the rail infrastructure proposed.
- State of readiness from competing taxi and bus infrastructure and degree of market sales share likely.
- Assessment of travel times for all comparative modes of transport during normal and peak times.

Promotion of rail services over conventional modes of transport:

- Rail services should aim to attract staff and the travelling public by providing both cost-effective and convenient travel to and from airport facilities through the operational day and night period.

Integrated action – designers should provide rail facilities that:

- Have the capacity with further investment in some cases to meet the operational requirements of the airport for the next 30 years.
- Meet the needs of the passengers and the local community on opening.
- Offer in-town or remote hotel check-in co-ordination, providing mechanisms, systems and railway carriages dedicated for moving and handling check-in baggage and hand cabin-sized baggage.
- Design systems that interact with one another, thereby providing passengers with seamless transition from the rail system to the airport environment.

5.2.2 Recommendations, intermodality and airport access

Intermodality strategy:

- Airport planners and operators should consider the provision of co-ordinated intermodality strategy plans. These should present the opportunity to reduce normal road traffic by no less than 10% if implemented successfully, which should be the objective.

REFERENCE

International Air Transport Association (2004). *Airport Development Reference Manual*, 9th edition. IATA.

5.3 BAA

The BAA (the former British Airports Authority) wants to make Heathrow the world's leading surface transport hub as well as its leading aviation hub. Its five-year Surface Access Strategy gives many detailed commitments, one overall aim of which is to raise the proportion of air passengers using public transport from 33% to 50%.

This strategy is a prime example of the initiative advocated by the IATA, and also represents a strong response to the Airports Council International commitment 4 of 2001 – to work closely with partners to develop and improve public transport (see Chapter 7).

REFERENCE

BAA Heathrow (2003). Shaping the vision: a surface access strategy for Heathrow. Progress Report, October.

5.4 Royal Institute of Chartered Surveyors, UK (RICS)

The study 'Transport Development Areas' (TDAs), supplemented by a guide to good practice for transport interchanges, envisages:

- A virtuous circle of land use plus transport planning leading to better integration.
- More sustainable development.
- Transfer of journeys/trips to more sustainable modes.
- Provision of key focus for spatial development frameworks and locational planning.
- A shared vision or strategy supporting the urban renaissance and delivering suitable outcomes on the ground.
- A degree of certainty regarding core densities – particularly residential – to be expected within a designated TDA.

REFERENCE

Royal Institute of Chartered Surveyors (2002). *Transport Development Areas, Guide to Good Practice*. RICS.

5.5 Chartered Institute of Logistics and Transport (CILT)

The report 'Passenger interchanges: a practical way of achieving passenger transport integration' contained 64 recommendations, including:

- Interchange should form part of highway authorities' Traffic Management Programmes, improving access to, and the circulation of, public transport services.
- Train operating companies could be encouraged to assess their stations to establish their potential for bus interchange.
- Bus operators should reappraise their bus route and terminal points to determine if potential exists for integrating their services better with rail or other bus operation.

REFERENCE

Institute of Logistics and Transport (1999). *Passenger interchanges: a practical way of achieving passenger transport integration*. ILT.

5.6 Transport for London

The report 'Intermodal transport interchange for London' (2001) gives design guidance, operational guidance and advice for partnership and consultation, and is found at: www.transportforlondon.gov.uk/tfl/reports_library_interchange.shtml.

REFERENCE

Transport for London (2001). *Intermodal transport interchange for London*. TfL.

5.7 Nottingham University, UK

Nottingham University maintains a database of references, found at www.nottingham.ac.uk/sbe/planbiblios/bibs/sustrav/refs/st11a.html.

5.8 National Center for Intermodal Transportation, USA (NCIT)

The NCIT was founded in 1998 as a University Transportation Center sponsored by the United States Department of Transportation. NCIT is a major national resource for educational, research and technology transfer activities involving intermodal transportation. The NCIT and its studies is a collaborative partnership between two universities, the University of Denver and Mississippi State University, and multiple disciplines within each university including business, law, engineering and science.

From single modal perspectives, the United States has developed one of the best transportation systems in the world. However, because each mode of transportation evolved independently of the others, they are not well integrated. As a result, it is difficult to transfer passengers and freight from one mode to another. Furthermore, some modes are overused, creating delays and hazards, while other modes are underused and have excess capacity. The NCIT believes that the overall contribution of the national transportation system can be increased by the creation of an intermodal system based on a more balanced and rational use of all modes of transportation. As such, the theme of the NCIT is the assessment, planning and design of the nation's intermodal transportation system with a focus on improving the efficiency and the safety of services for both passengers and freight by identifying ways to better utilise the strengths of the individual modes of transportation.

What about a comprehensive approach to reclamation of the transport interchange, social, commercial and sustainable?

6 Taxonomy of rail, bus/coach and air transport interchanges

Integral structures/linked and contiguous structures, vertical and horizontal separation are considered.

Terminology used for rail services at airports is consistent with the types used in the IATA *Airport Development Reference Manual*, 9th edition.

GENERAL REFERENCES FOR CHAPTER 6
Civil Aviation Authority (CAA) data (UK traffic figures).
Airports Council International (ACI) website (worldwide traffic figures).

6.1 Airport/railway interchange: vertical separation

6.1.1 Zurich Airport, Switzerland

Interchange – national railway station below terminal building, connected by new Landside Centre.

For over 30 years, Zurich Airport has boasted integration of air travel with the European rail network, by the location beneath the landside complex of the airport of a through rail station. One step to enhance the modal interchange experience was the experimental introduction in the late 1970s of baggage trolleys that could be safely taken on the escalators delving to the subterranean rail station. Measured against the airport's passenger handling capacity 30 years ago of 6 million passengers per year, its accessibility by rail was exceptional. The airport now handles well over 20 million passengers per year.

Trains make the 10-minute trip to Zurich main station every 10 minutes. Public transport is used by 52% of passengers and 20% of airport employees.

New piers in 1975, 1985 and 2003 have increased capacity, which will reach 35 million by 2010. The second new pier (A) has the form of three piers or lounges laid end to end. Each unit has its own security control. The outer two are reached from the stem of the pier by moving walkways that 'duck under' into a mezzanine level, which also provides a segregated route for arriving passengers. The third pier is in fact a midfield satellite 'dock' with up to 27 stands, including two for Code F (A380) aircraft, connected by a cable-driven people-mover, the Skymetro, with its station beneath the new Airside Centre. Two more key projects completing in 2003 and 2004 respectively are the Landside Centre and Airside Centre.

TEAM, LANDSIDE CENTRE AND AIRSIDE CENTRE

CLIENT: Unique (Flughafen Zurich AG)
ARCHITECTS: British Nicholas Grimshaw & Partners with Swiss Itten and Brechbuhl AG
ENGINEERS: British Arup with Swiss Ernst Basler & Partners

Transport terminals and modal interchanges

The Landside Centre unites the access to the rail station with the landside road system and the landside terminal concourses under a sophisticated roof between two car park structures.

The Airside Centre, by the same team, unites the two terminals with a retail complex and passport control. The centre has three levels, the Skymetro station, the baggage reclaim/arrivals hall and uppermost, under a curved 250-metre-long steel roof, lounge areas and plentiful retail.

A study by Hamburg-Harburg Technical University in 2003 focused on the landside accessibility of Zurich Airport, demonstrating how rail service replaces feeder flights, extends the catchment area, relieves traffic on surrounding roads and increases reliability of access time. Noting the present modal split of 50% by public transport, not only an aim but a condition of the operation of the airport, the study makes specific recommendations to enhance rail accessibility. This is measured by increasing the population

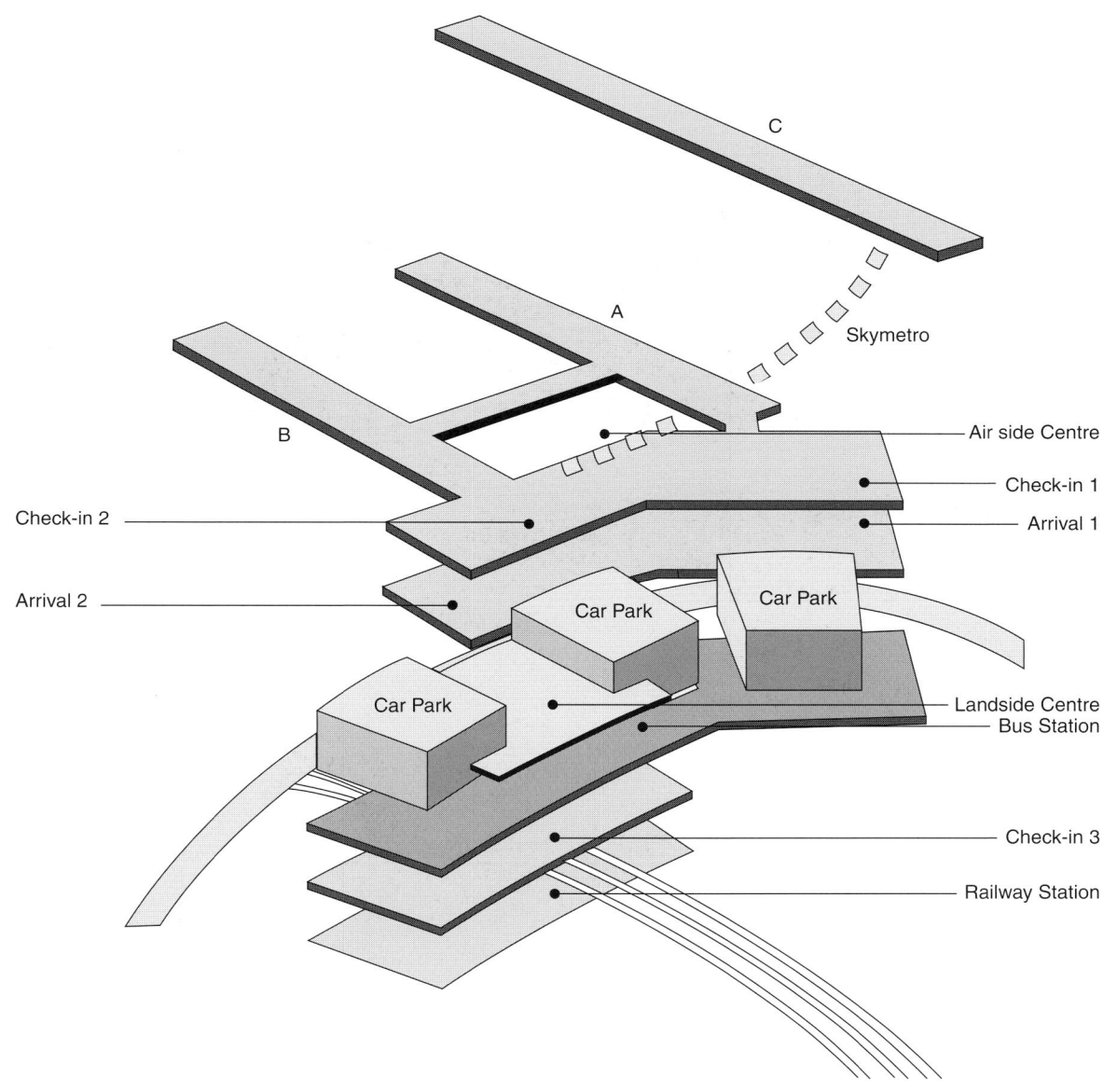

6.1 Zurich Airport: diagram of terminals based on Unique Airport website.

Airport/railway interchange: vertical separation

6.2 Interior of Landside Centre (Courtesy of Nicholas Grimshaw & Partners).

within one hour from approximately 1.7 million as at present to 3 million or even 4 million by 2020 if Zurich and Euro Airport (Basel-Mulhouse) are linked by high-speed train.

REFERENCES

Arnet, O. and Brunner, A. (2002). Swiss bliss. *Passenger Terminal World Annual Technology Showcase Issue*.

Littlefield, D. (2003). Inner beauty. *Building Design*, 22 August.

Wagner, T. (2003). Landside Accessibility Report. European Centre of Transport and Logistics, TU Hamburg-Harburg.

6.3 Roof of Landside Centre (Courtesy of Nicholas Grimshaw & Partners).

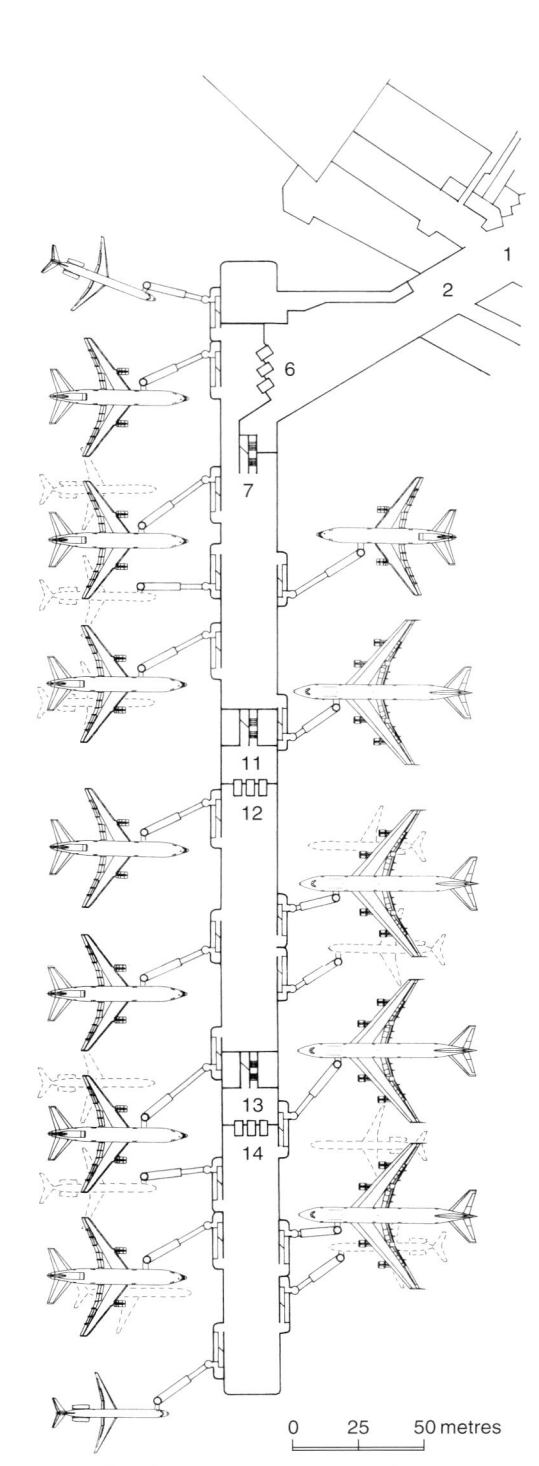

▲ 6.4 Pier A plan, main passenger level.

▲ 6.5 Pier A view (courtesy of Thyssen Henschel).

KEY TO ZURICH PIER A PLAN AND SECTION

1. Terminal A
2. Outbound immigration control
3. Inbound immigration control
4. Transit
5. Baggage sorting area
6. Security control, gate zone 1
7. Gate zone 1
8. Arrivals
9. Departures
10. Loading bridge
11. Security control, gate zone 2
12. Gate zone 2
13. Security control, gate zone 3
14. Gate zone 3

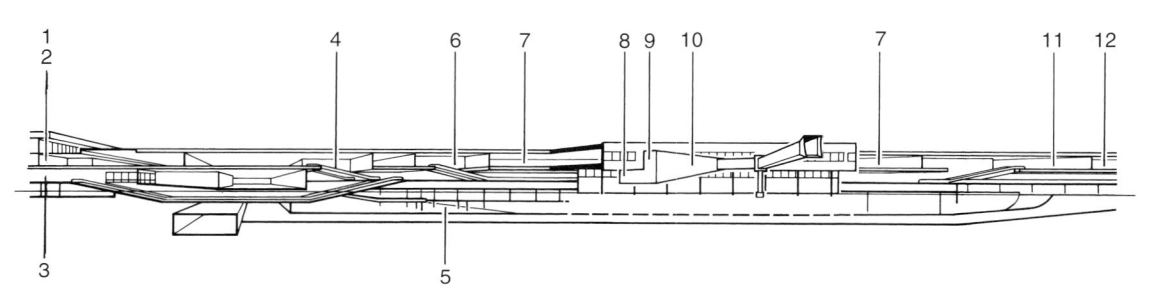

▲ 6.6 Pier A section (courtesy of Flughafen Zurich Informationsdienst).

6.1.2 Amsterdam Schiphol Airport, The Netherlands

Interchange – national railway station below terminal building, combined with buses and taxis at ground level.

Schiphol Plaza is the modal node between the underground train station of the national railway network, the car park and arriving vehicle lanes for taxis, buses and private cars on landside, and the terminals on airside, so it is the centre of all activity. Trains serve the city of Amsterdam 16 minutes away and Rotterdam is only 45 minutes away.

Schiphol Airport's Amsterdam site was first chosen for an airfield in 1919, on a reclaimed polder with the historical name of Schipshol (hell for shipping). Originally in the sea lanes 16 kilometres south of the city of Amsterdam, Schiphol was the point of departure for the first commercial flight from Amsterdam to London Croydon by the British airline Air Transport and Travel. The same airline was the operator of the first scheduled service of all, from London Hounslow to Paris in 1919, as described in Chapter 2. For 20 years, until destruction in 1940, Schiphol was the base both for the Dutch airline KLM and for the manufacture of Fokker aircraft. By 1938 it had become only the second airport in Europe with paved runways and was handling over 10 000 passengers per year.

Post-war rebuilding was led by the reconstruction and lengthening of the runways, followed by new terminal buildings. The following chronology charts the post-war development of Schiphol Airport, using present pier designations:

Post-war operations recommence	1945
Tangential runway master plan	1949
Two 3300 m runways open	1960
Terminal with C, D and E piers and two more runways, with 25 aircraft stands and capacity of 4 million passengers per year	1967
C Pier extended, nine more aircraft stands and total capacity of 8 million passengers	1971
D Pier modernised	1974
New F pier, eight wide-body stands and terminal extended, with total capacity of 18 million passengers per year (8 million handled)	1975
Railway link opened	1978
Ten million passengers per year mark passed	1980
E Pier rebuilt with 10 wide-body stands	1987
D Pier extension with 13 more MD80 or B737 stands and 16 million passengers per year handled	1990
Five-year plan raising capacity to 27 million passengers (terminal extension and G Pier)	1993
Further plan to raise quality and service standards (B Pier and D Pier 'alternate' northern extension)	1995

Prior to 1995 the strategy was to provide a single terminal complex, offering convenient transfer for the 37% of passengers who then changed planes at Schiphol. Then, with the splitting of intra-European Community traffic, treated as domestic traffic, from truly international traffic, a contiguous terminal extension was demanded. The pre-existing terminal would be dedicated to EC traffic and the extension to non-EC traffic, with the full range of customs and immigration facilities.

The northern extension to D Pier, with a second circulation level, provides for alternative use between EC and non-EC, obviating the need for repositioning of aircraft.

▲ 6.7 Schiphol Airport, aerial view (Courtesy of NACO BV).

KEY TO AERIAL PHOTOGRAPH (FIGURE 6.7)

B   B Pier (regional airlines), 1994
C   C Pier, 1967 with 1971 A-head extension
D   D Pier, 1967
DS   D Pier southern extension, 1990
DN D Pier northern extension, 1995
E   E Pier, 1967; rebuilt 1987, extended 2000
F   F Pier, 1975
G   G Pier, 1993
T1   Original terminal
T2   Terminal expansion
P   Schiphol Plaza (see Figures 6.8 and 6.9)

6.8 Schiphol Plaza, interior (Courtesy of NACO BV).

6.9 Schiphol Plaza, exterior (Courtesy of NACO BV).

The airport now handles 40 million passengers per year (ACI 2003 figure), and is seen as a prime example of an Airport City.

TEAM, SCHIPHOL PLAZA

ARCHITECTS: Benthem Crouwel Architekten in association with NACO, Netherlands Airport Consultants

6.1.3 Vienna Airport, Austria

Interchange – dedicated high-speed train and urban railway terminus below terminal building.

Vienna Airport, being the easternmost of the European Union up to 2004, has developed a hub for Eastern Europe and handled 13 million passengers in 2003, of which over 35% were transfers.

A key feature of Vienna International Airport's strategy is the introduction of the City Airport Train (CAT), a joint venture with the Austrian Railroad Company. Trains provide a 16-minute service from the city centre with in-town check-in possible.

The new projected Skylink North-East extension to the terminal area will be a partly four-storey building, including separate levels for baggage handling, Schengen, non-Schengen and transfer passengers. The cost is Euro 400 million and the construction will take place between 2005 and 2007. A below-ground link to the new railway station is also an essential part of the plan.

TEAM, SKYLINK

CLIENT: Vienna International Airport
DESIGN TEAM: Fritsch, Chiari & Partner, ZT GmbH, Vienna

6.10 Vienna Airport: existing aerial photo – CAT station below landside terminal (Courtesy of Vienna International Airport).

▲ 6.11 Diagram of future VIE Skylink (Courtesy of Vienna International Airport).

REFERENCE

Tmej, M. and Mayerhofer, P. (2003). Presentation at Future Terminal Conference, London.

6.1.4 Heathrow Airport Terminal 5, UK

Interchange – metro terminus below terminal building together with dedicated Heathrow Express city centre service terminus.

The rail (and bus) interchange at Heathrow Terminal 5 has been designed from the start ready for completion in 2008. While bus/coach and rail usage at Heathrow in 2002 was 34%, the aim is to reach 40% by 2008, aided by the improved interchange at T5, and the longer-term target is 50%.

The terminal building has a new road traffic interchange with the M25 motorway and a new spur takes the London Underground Piccadilly Line to Terminal 5. The new terminus of the Heathrow Express is designed in such a way that, in future, trains can run through to the west, connecting with a projected link to Staines to the south and a possible link to Slough to the north-west of the airport. The first of these, titled Airtrack, offers links to the whole conurbation of south-west London and the counties of Surrey and

▲ 6.12 Heathrow Terminal 5: artist's impression of Interchange zone on west face of main terminal, linking terminal to car park, buses, coaches and rail systems (Courtesy of BAA Heathrow Terminal 5 website).

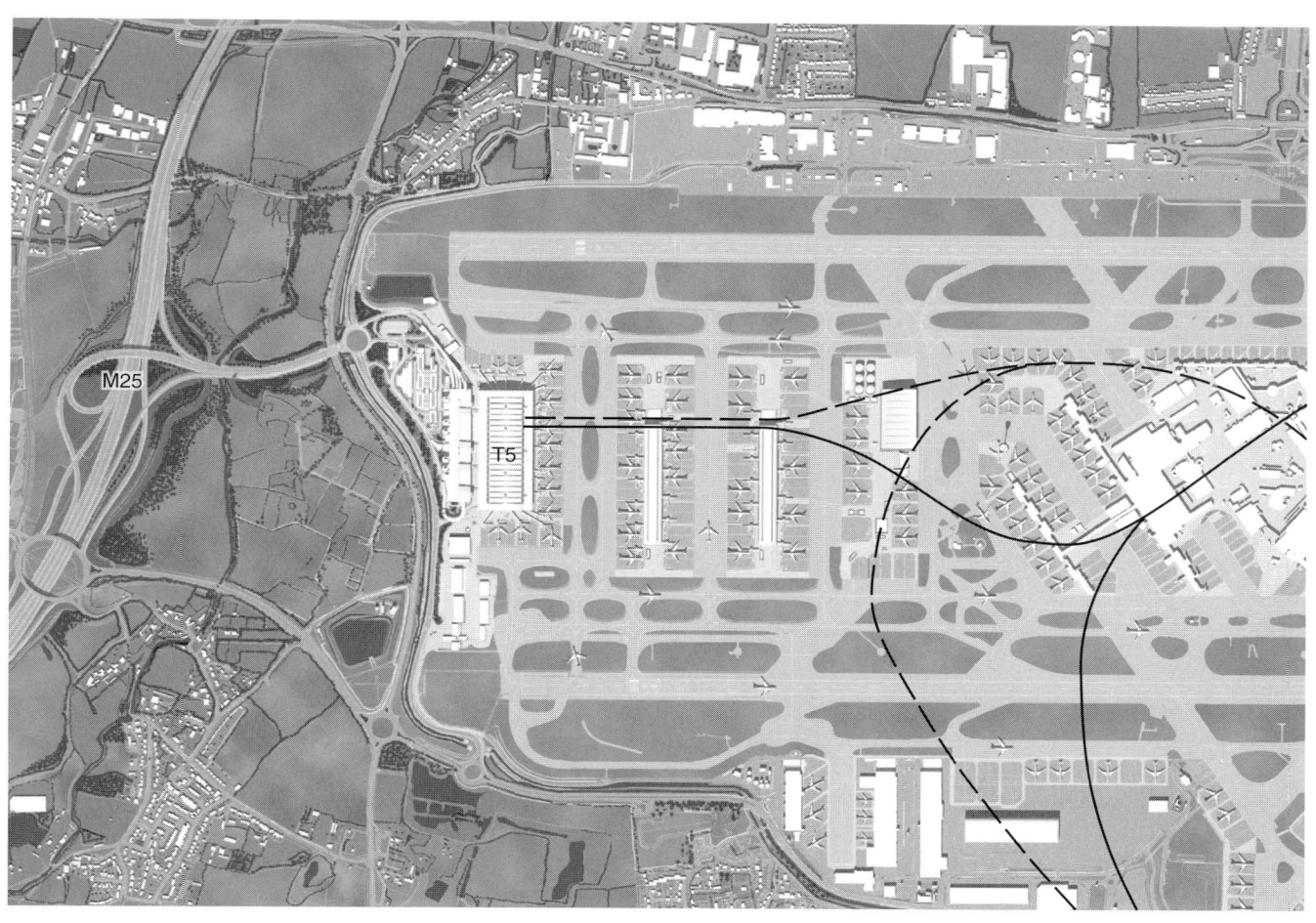

Transport terminals and modal interchanges

▲ 6.13 Aerial view, with underground Heathrow Express shown as a continuous line and Piccadilly Line as a dotted line (Courtesy of BAA Heathrow Terminal 5 website).

Hampshire. The second of these would offer main-line rail services to the whole of the Thames Valley and the West of England. An Interchange zone on the west or landside face of the main terminal building connects to buses, coaches and car parking.

Heathrow Terminal 5 itself, made possible by the clearance of a redundant sewerage facility to the west of the original terminal area, has been the subject of a long gestation period. When complete in 2008, the first phase of the facility, with a central terminal and one linear pier or satellite-transverse between the runways and linked by a sub-apron airside people-mover, will have 42 aircraft stands and a capacity of 20 million passengers per year. This traffic is expected to be split in the ratio 4:6:10 between domestic, short haul and long haul, with an arrivals and departures busy hour rate of 3300 passengers. The second phase, expected to be complete in 2011, will add two more satellites and raise the annual capacity to 30 million passengers per year. This traffic is expected to be split in the ratio 5:10:15 between domestic, short haul and long haul, with an arrivals busy hour rate of 3850 passengers and departures busy hour rate of 4200 passengers. Detailed design has been geared to the requirements of British Airways.

Airport/railway interchange: vertical separation

▲ 6.14 View of main terminal from the south-east (Courtesy of BAA Heathrow Terminal 5 website).

Maximum future-proof design has led to safeguarding the underground transit system which takes passengers from the main terminal to island satellites for future extension to the Central Terminal Area of Heathrow.

TEAM

CLIENT:	BAA plc
PRINCIPAL ARCHITECT, APPOINTED 1989:	Richard Rogers Partnership
STRUCTURAL ENGINEER:	Arup

REFERENCES

BAA Heathrow (2003). Shaping the vision: a surface access strategy for Heathrow. Progress Report, October.

Berry, J. (2003). Presentation at Future Terminal Conference, London.

6.1.5 Heathrow Airport Terminal 4, UK

Interchange – metro station (one-way) below terminal building. Note also the dedicated Heathrow Express terminus and bus stands in forecourt.

This terminal was opened, as British Airways intercontinental terminal, in 1986, and a pier extension added in the early 1990s in response to improved passenger service standards requiring loading bridge access to aircraft in preference to buses. In spite of severe height restrictions operative at the time of its design, the terminal offered complete vertical segregation of arriving and departing passengers and an uninterrupted 650-metre-long airside departures concourse.

Thirteen per cent of passengers use the tube (CAA 2002 figure). The extension in 1986 of the Piccadilly Line to form a clockwise single-track loop between Hatton Cross and the Central Area improved the service considerably.

Nine per cent of passengers use the Heathrow Express, which provides a 15-minute service to Paddington via Heathrow Central every 15 minutes (CAA 2002 figure).

TEAM

CLIENT: BAA plc, formerly British Airports Authority
ARCHITECTS: Scott Brownrigg Ltd
STRUCTURAL ENGINEER: Scott Wilson

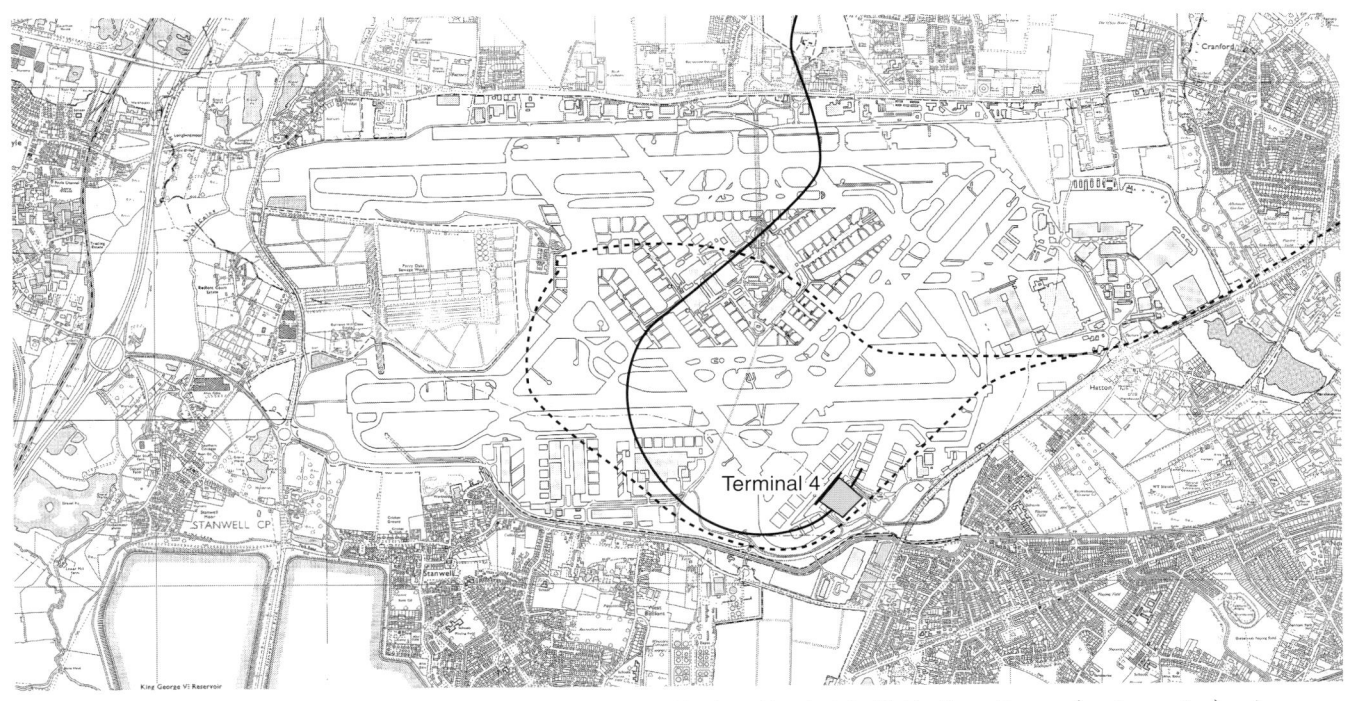

6.15 Heathrow Terminal 4 with Heathrow Express (continuous line) and Piccadilly Line (dotted line) approach routes (courtesy of Heathrow Airport Ltd).

Transport terminals and modal interchanges

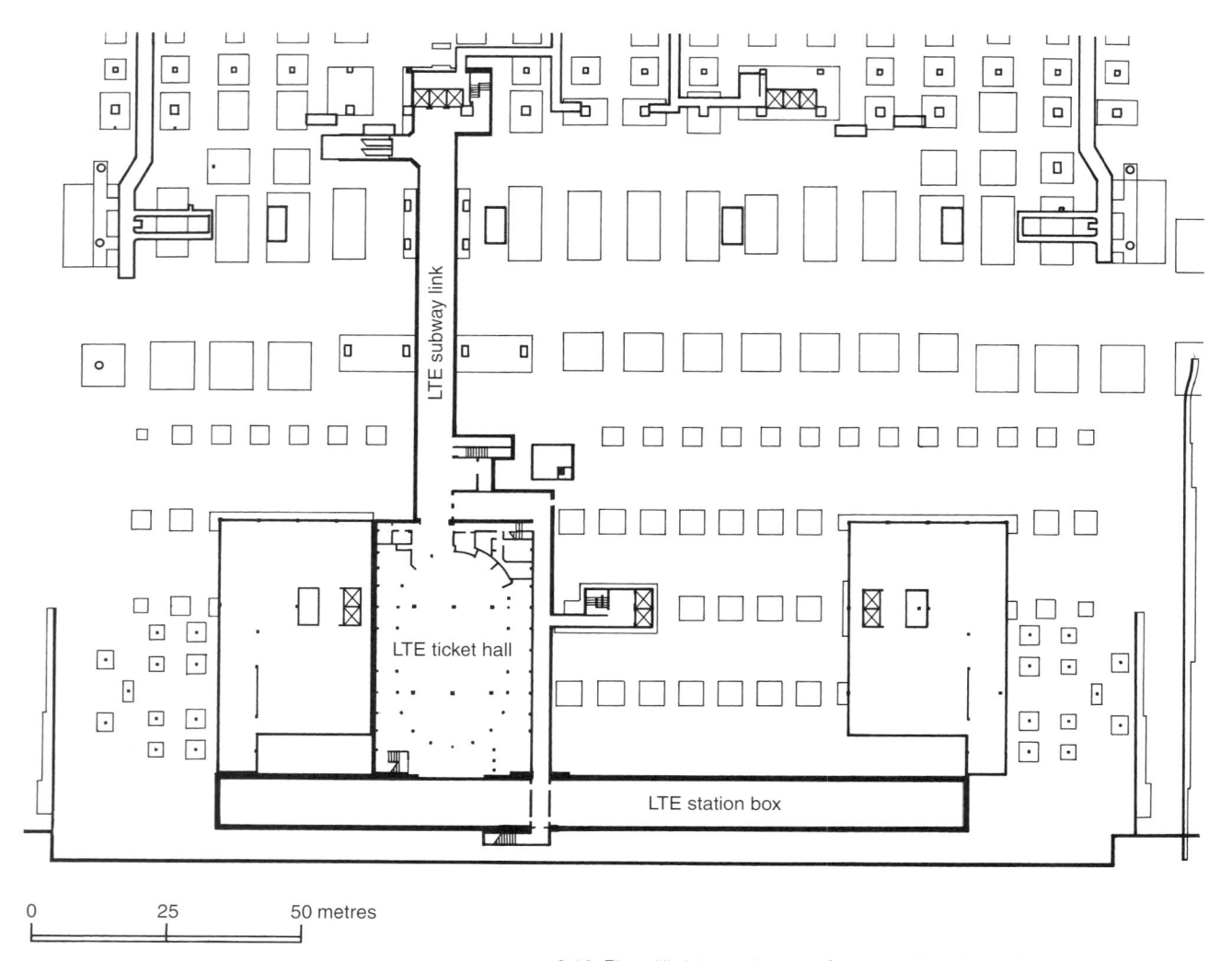

6.16 Piccadilly Line station plan (courtesy of architects Scott Brownrigg and London Underground Ltd).

6.17 London Underground station interior.

52   Airport/railway interchange: vertical separation

Transport terminals and modal interchanges

▶ 6.18 Arrivals forecourt bus stands.

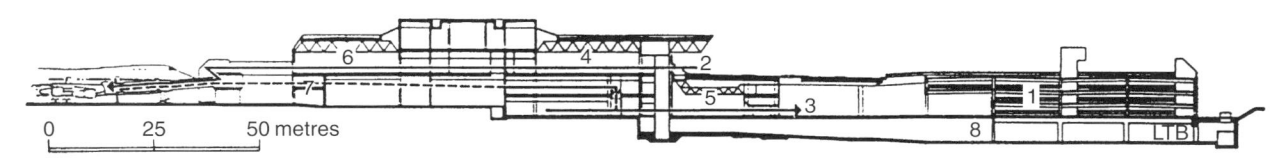

▲ 6.19 Arrivals forecourt plan (Courtesy of architects Scott Brownrigg). Key: 1 Multi-storey car park; 2 Departures forecourt; 3 Arrivals forecourt; 4 Departures concourse; 5 Arrivals concourse; 6 Airside concourse; 7 Arrivals corridor; 8 London Underground station.

Airport/railway interchange: vertical separation **53**

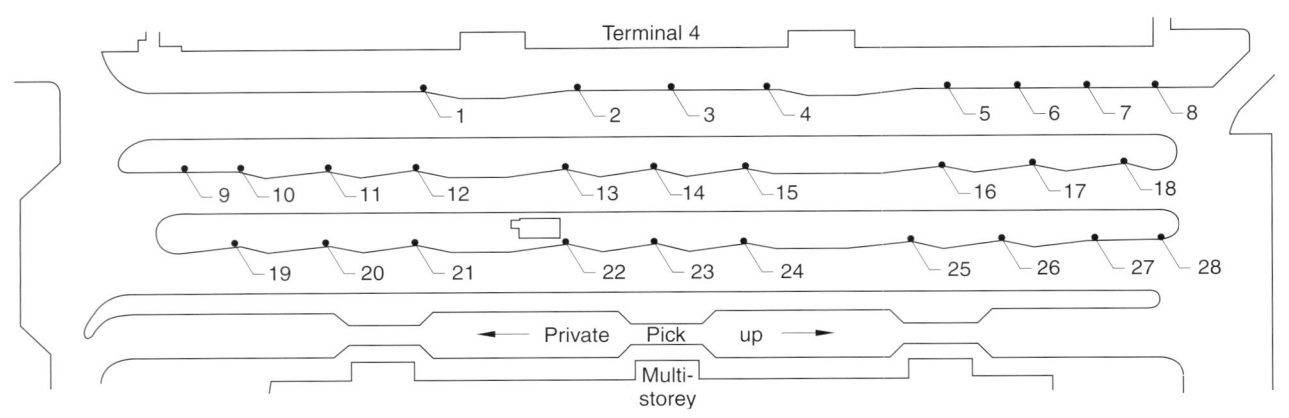

6.20 Terminal cross-section (Courtesy of architects Scott Brownrigg). Key: 2–28 represent bus bays. 1 Taxis; 2 Railair/Rickards/Alder Valley; 3 Long-term car park; 4 Car rental concession; 5 Transfer to Terminal 1; 6 Transfer to T2 and T3; 7 Alder Valley/Careline; 8 Staff car park; 9 Jetlink; 10 National Express; 11 City of Oxford; 12 Speed link; 13 Flightline; 14 Flightline/Airbus; 15 Airbus; 16 Flightlink; 17 National Express; 18 Southend/Premier; 19–24 Group travel; 25 Green line; 26 London Country; 27 Local hotels; 28 Off-airport car parking/rental.

6.21 Check-in hall (Courtesy of architects Scott Brownrigg).

REFERENCE

BAA Heathrow (2003). Shaping the vision: a surface access strategy for Heathrow. Progress Report, October.

Transport terminals and modal interchanges

6.1.6 Chicago O'Hare Airport, USA

Interchange – metro terminus below three out of four terminal buildings and linked to the fourth by people-mover.

The second busiest (with 69 million passengers in 2003), if not the largest, airport in the world can be regarded as a 'transfer station' between several modes of transport. The three domestic terminals, numbered 1–3, are served by a vast car parking building with the terminus rail station of the CTA Blue Line below. This line offers a 45-minute service at 8-minute intervals to the city centre, and carries 4% of all passengers travelling to and from the airport.

Terminal 1, built for United Airlines in 1988, has two 480-m-long concourses with soaring glazed vaults. One is adjacent to the two-level forecourt and provides immediate walk-on access to 25 aircraft stands. The other is reached through a sub-apron concourse and has the form of an island satellite with 27 stands. A baggage handling area of 7500 m² is located under the apron alongside the subway linking the two concourses.

TEAM, TERMINAL 1 AND CTA STATION UPGRADE

ARCHITECTS: Murphy & Jahn

▼ 6.22 Chicago O'Hare Airport: view of rail station beneath parking building serving Terminals 1, 2 and 3 (courtesy of Department of Aviation, City of Chicago).

Airport/railway interchange: vertical separation **55**

6.23 Terminal 1 (United Airlines) passenger tunnel (Courtesy of architects Murphy & Jahn, New York).

6.24 Departures concourse (Courtesy of architects Murphy & Jahn, New York).

Transport terminals and modal interchanges

▲ 6.25 Check-in hall (Courtesy of architects Murphy & Jahn, New York).

Terminal 5 for international traffic is not served directly by the CTA train, but is connected by a people-mover, which also stops separately at Terminals 1, 2 and 3. The layout of this terminal is determined by the available land and road access. The new requirements of the Federal Inspection Service for an international terminal dictated total centralisation and segregation of passenger movement. The terminal is designed to handle all international arrivals, but only departures by foreign airlines: this is demonstrated by the design capacities for arrivals and departures of 4000 and 2500 passengers per hour respectively. Dominant features of the building are the great arching roof over the 250-m-long ticketing pavilion, the Galleria link to the airside and the single-side three-level piers with a total length of 500 m.

TEAM, TERMINAL 5

ARCHITECTS: Perkins & Will

Airport/railway interchange: vertical separation

▲ 6.26 Terminal 1 aerial view (Courtesy of architects Murphy & Jahn, New York).

KEY TO TERMINAL 5 SECTION AND PLANS (FIGURES 6.28–6.30)

1 People-mover
2 Ticketing pavilion
3 Galleria
4 Departures security
5 Departures corridor
6 Arrivals corridor
7 Passport control
8 Baggage reclaim
9 Customs inspection
10 Arrivals concourse

Transport terminals and modal interchanges

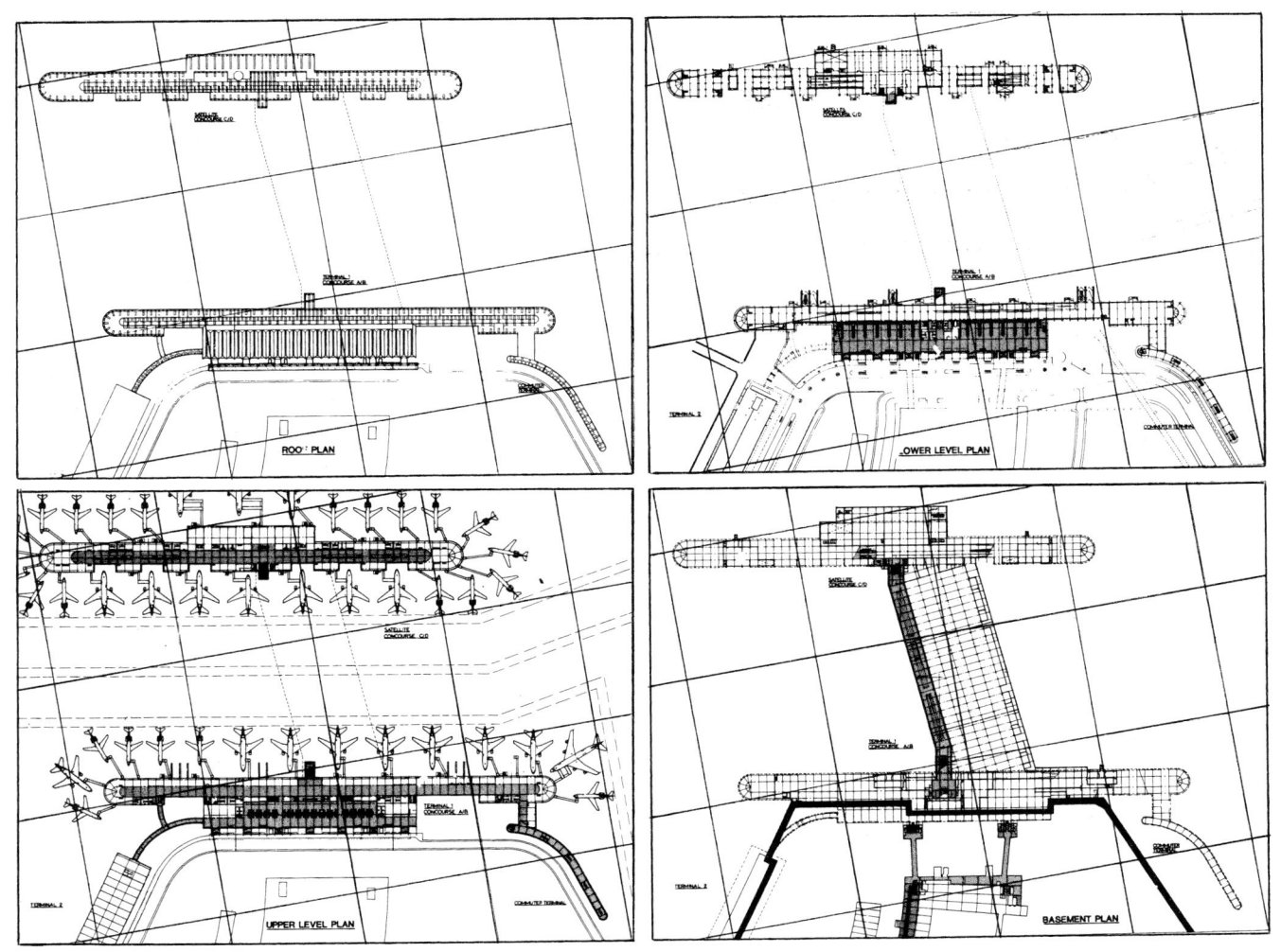

▲ 6.27 Terminal 1 plans (Courtesy of architects Murphy & Jahn, New York).

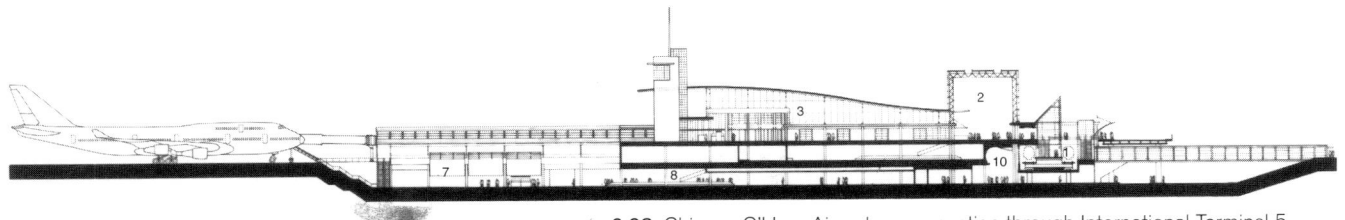

▲ 6.28 Chicago O'Hare Airport: cross-section through International Terminal 5 (Courtesy of architects Perkins & Will, Chicago).

Airport/railway interchange: vertical separation **59**

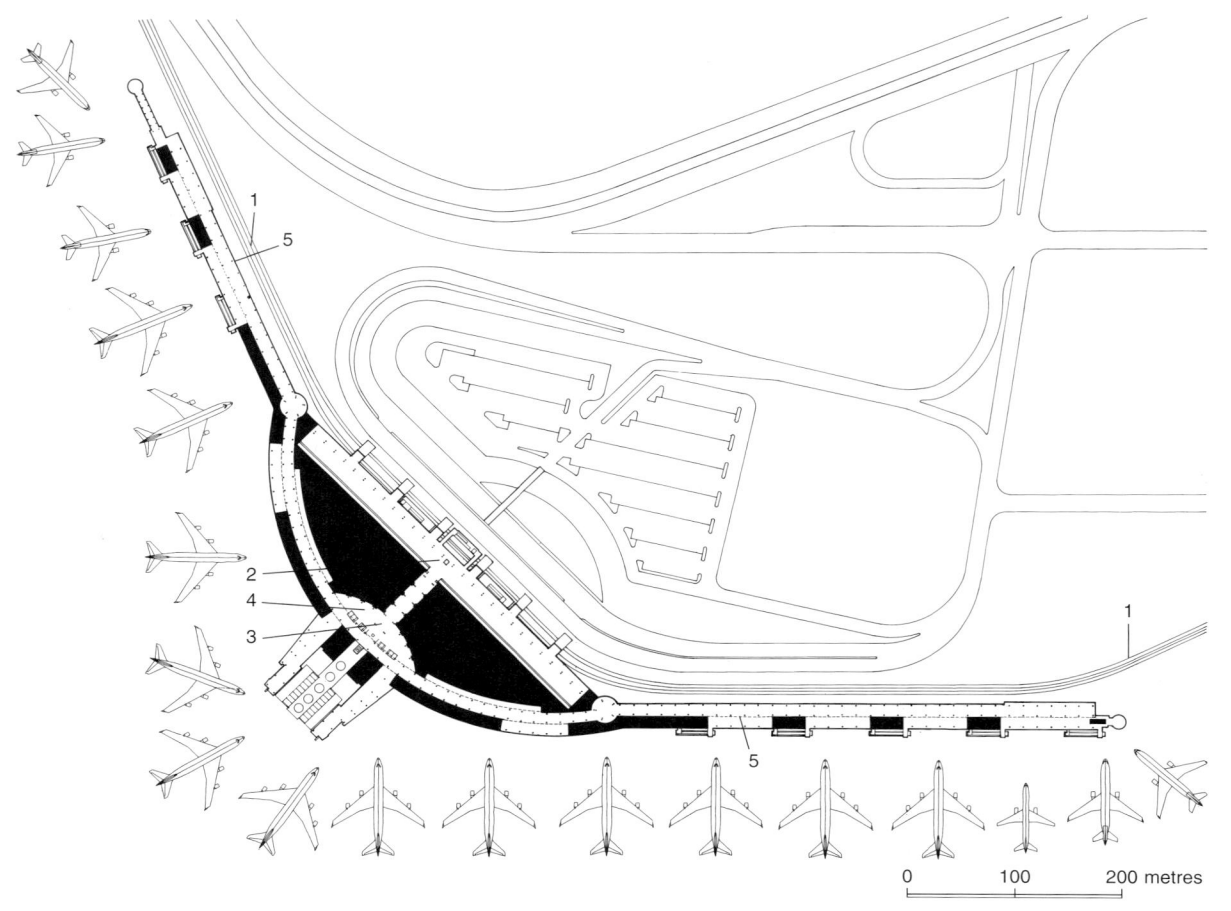

▲ 6.29 Terminal 5 upper level plan (Courtesy of architects Perkins & Will, Chicago).

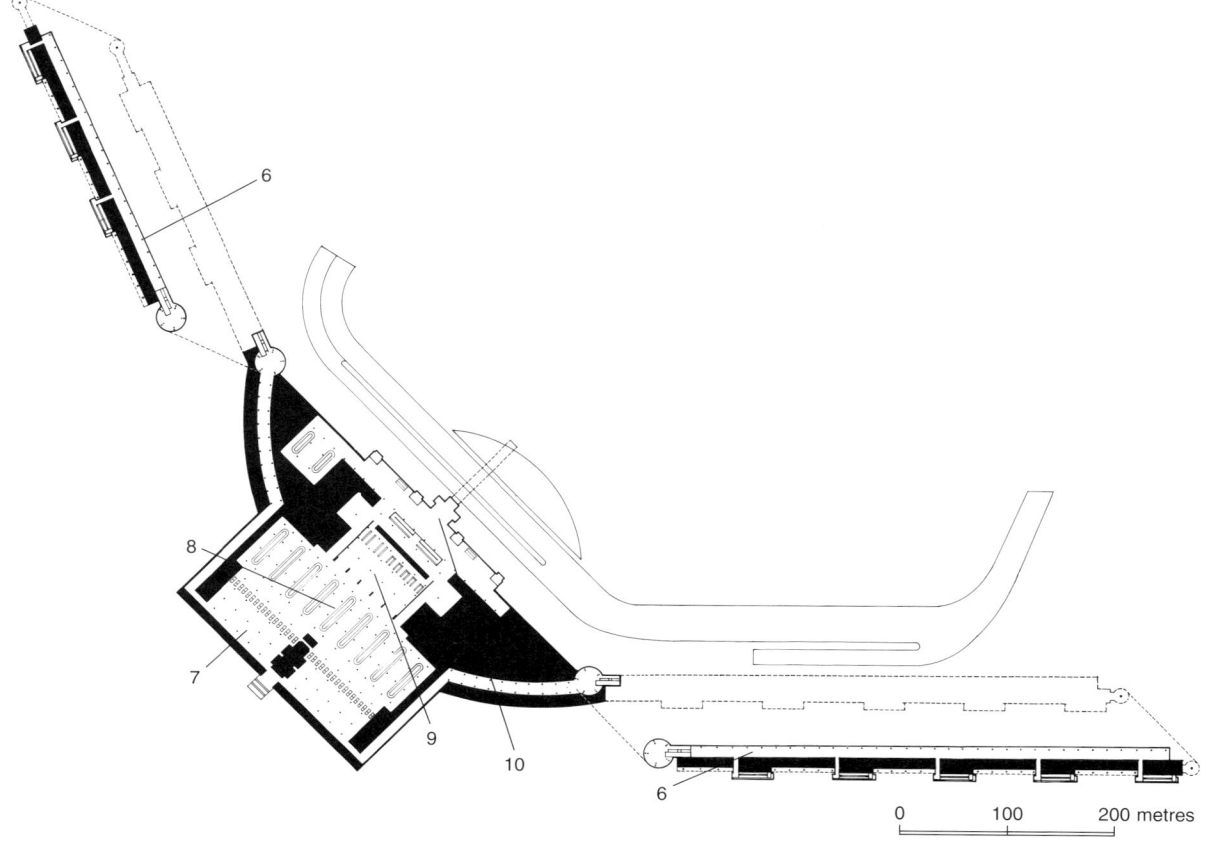

▲ 6.30 Terminal 5 lower level plan (Courtesy of architects Perkins & Will, Chicago).

▶ **6.31** Exterior view (Courtesy of architects Perkins & Will, Chicago).

▶ **6.32** Interior view (Courtesy of architects Perkins & Will, Chicago).

6.2 Airport/railway interchange: contiguous

6.2.1 Paris Charles de Gaulle Airport, France

Interchange – national and high-speed railway station between six out of seven terminal buildings.

This airport has developed over a period of 30 years since the first terminal opened. The famous hollow-drum-shaped terminal is now quite separate from the parts of Terminal 2.

Terminal 2 comprises six modular terminals built over a period of over 20 years. The first four are Terminals 2A to 2D, each with approximately six contact gates, and these were followed by the contiguous railway station surmounted by a hotel. The rail tracks at low level intersect the axis of the terminal buildings. Two more terminals, numbered 2E and 2F, have been developed beyond the rail station.

An interesting feature of Terminal 2C is the luffing ramps. The introduction of segregated passenger routes on different levels in the airside parts of the terminal leads to the need for passengers to descend ramps from and to each of these levels in order to enter the fixed end of a loading bridge. An economy measure is the use of a single ramp which pivots at the fixed end of the loading bridge, at a level approximately halfway between the two levels of the terminal. This moves on rails from one level to the other according to whether passengers are embarking or disembarking.

The airport now handles 48 million passengers per year (ACI 2003 figure).

Usage of public transport and the rail station, with both local services and high-speed long-distance services, is currently as follows:

Buses	14%
Local rail	16.5% – 35 minutes to central Paris, at 8- to 15-minute intervals
TGV	3% – Lyon and Lille lines with 25 trains per day.

▲ 6.33 Paris Charles de Gaulle Airport, aerial view, 2003 (Courtesy of Aeroports de Paris).

Transport terminals and modal interchanges

▲ 6.34 (a–e) Views of rail station and adjacent aircraft stands (Courtesy of Aeroports de Paris).

Airport/railway interchange: contiguous **63**

Transport terminals and modal interchanges

6.35 Terminal 1 aerial view, inflexible but intended to multiply (Courtesy of Aeroports de Paris).

6.36 Luffing bridge at Terminal 2C (Courtesy of Aeroports de Paris).

64 Airport/railway interchange: contiguous

6.2.2 Frankfurt Airport, Germany

Interchange – national and high-speed railway station adjacent to and regional rail station below one terminal building and linked to another by people-mover.

The airport now handles 48 million passengers per year (ACI 2003 figure).

This airport has developed two terminals, the first being linked to rail stations and hotels and the second, opened in 1994, with a people-mover link to the first. There will soon be a third terminal on the other side of the main runway.

The new AIRail centre, a high-tech office and hotel complex under construction above the national rail station, is being marketed as the most mobile workplace in Europe. Quite apart from the connections offered by the airport and airlines, the ICE Inter-City Express rail offers 96 arrivals and departures every day. It is a nine-storey complex with 185 000 m^2 of offices and two hotels – a horizontal skyscraper.

TEAM, AIRAIL

CLIENT: Fraport AG
ARCHITECTS: JSK International Architekten und Ingenieure GmbH

6.37 Frankfurt Airport: view of Terminal 1 (Courtesy of Flughafen Frankfurt am Main AG).

Transport terminals and modal interchanges

▶ 6.38 View of Terminal 2 (Courtesy of Flughafen Frankfurt am Main AG).

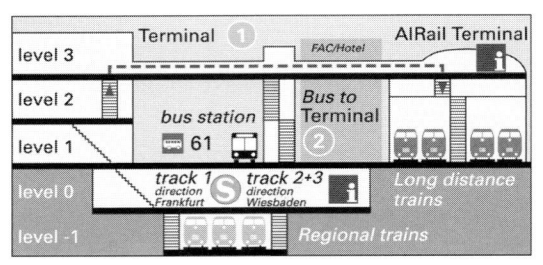

▲ 6.39 Cross-section showing regional rail station and national rail station/AIRail (Courtesy of airport website).

▶ 6.40 Diagrammatic plan showing regional rail station and national rail station/AIRail (Courtesy of airport website).

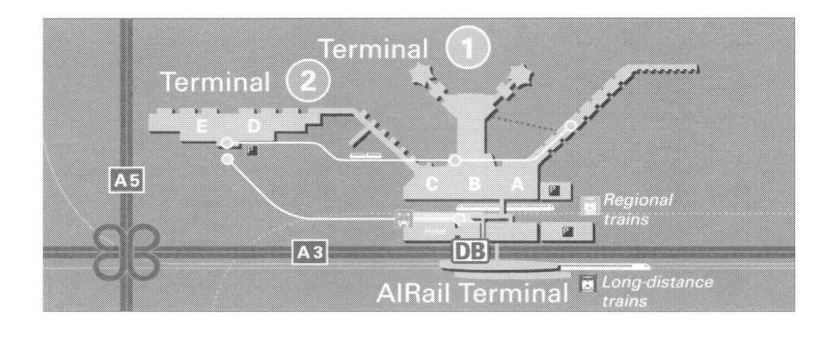

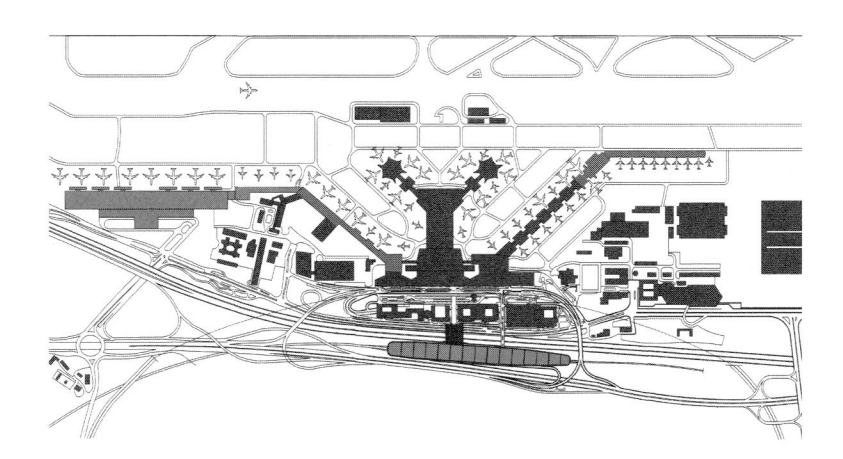

▶ 6.41 Detailed plan of airport terminals (Courtesy of JSK International Architekten und Ingenieure GmbH).

66 Airport/railway interchange: contiguous

Transport terminals and modal interchanges

▶ 6.42 Aerial view of Terminal 1 (without the new AIRail 'horizontal skyscraper' above the station) (Courtesy of Flughafen Frankfurt am Main AG and JSK International Architekten und Ingenieure GmbH).

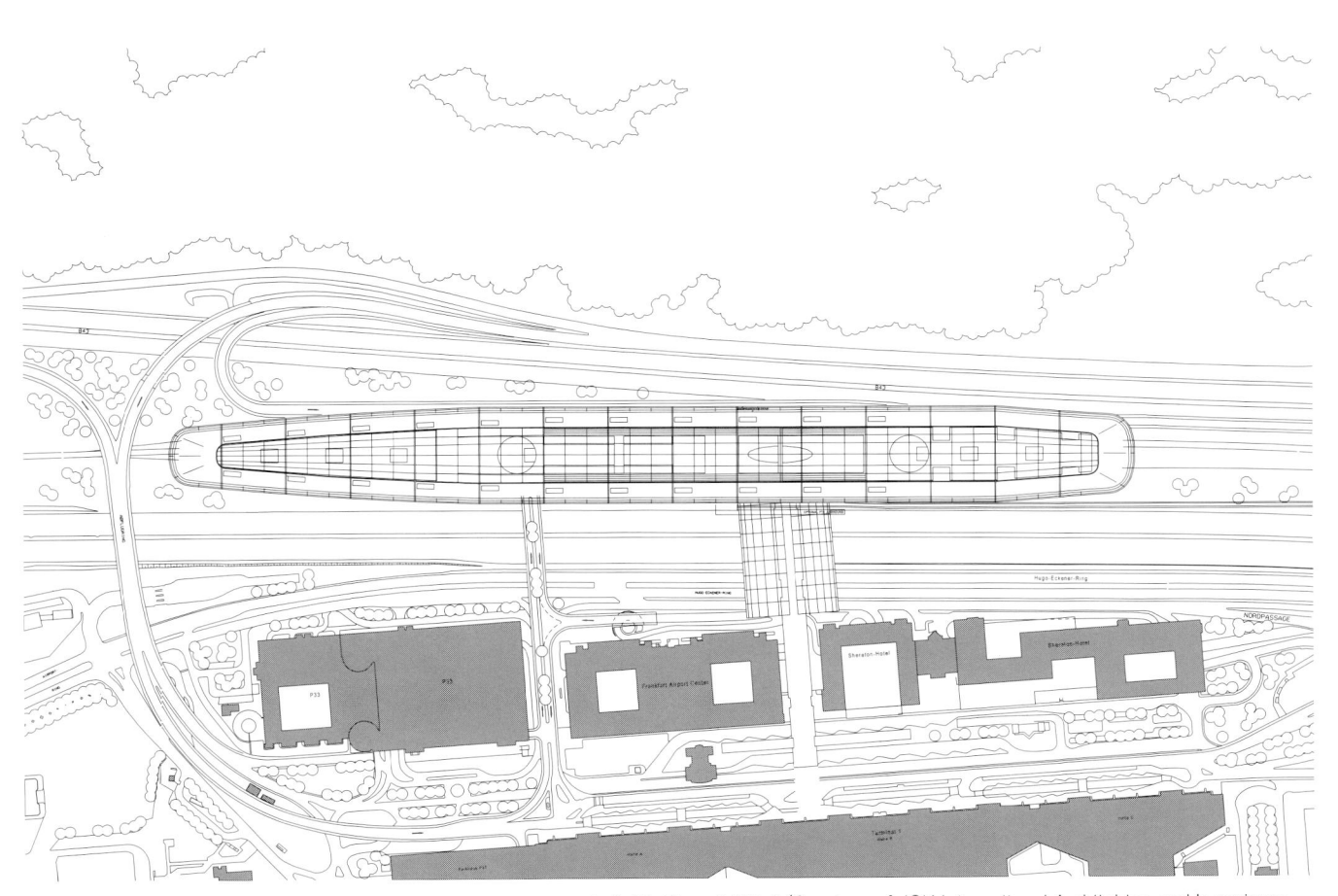

▲ 6.43 Plan of AIRail (Courtesy of JSK International Architekten und Ingenieure GmbH).

Airport/railway interchange: contiguous **67**

6.2.3 Stansted Airport, UK

Interchange – national railway terminus beside terminal building.

The airport handled 14 million departing and arriving passengers in 2002.

In 1991 this new terminal project increased to seven the number of terminals and to nearly 70 million the annual passenger handling capacity of London's three major airports. The central building is a sophisticated two-storey shed with all passenger functions at the upper level and all supporting facilities at the lower level, including a British Rail railway station link. The structural form gives large spans for maximum flexibility and highly disciplined building services. Each structural bay has a central rooflight fitted with reflectors suspended beneath the ceiling. Each 36-metre square structural bay is supported on a tree, which also contains all artificial lighting, information systems, air supply and extract, etc.

A rapid transit system running beneath the apron links the airside face of the terminal with two island satellites (each with 10 contact stands) and a third pier has been added with corridor link to the main terminal level. The terminal building itself has also been extended by the addition of two 36-metre-wide bays to the original five.

Beside the ground level of the terminal is the terminus of the Stansted Express, a dedicated rail service to Liverpool Street Station in the City of London, offering a 45-minute service every 15 minutes. This contributes to the achievement of 62% public transport usage to and from Central London. However, the overall public transport usage figures are 34% (28% by rail and 6% by bus) – figures from GLA Planning Report 2001.

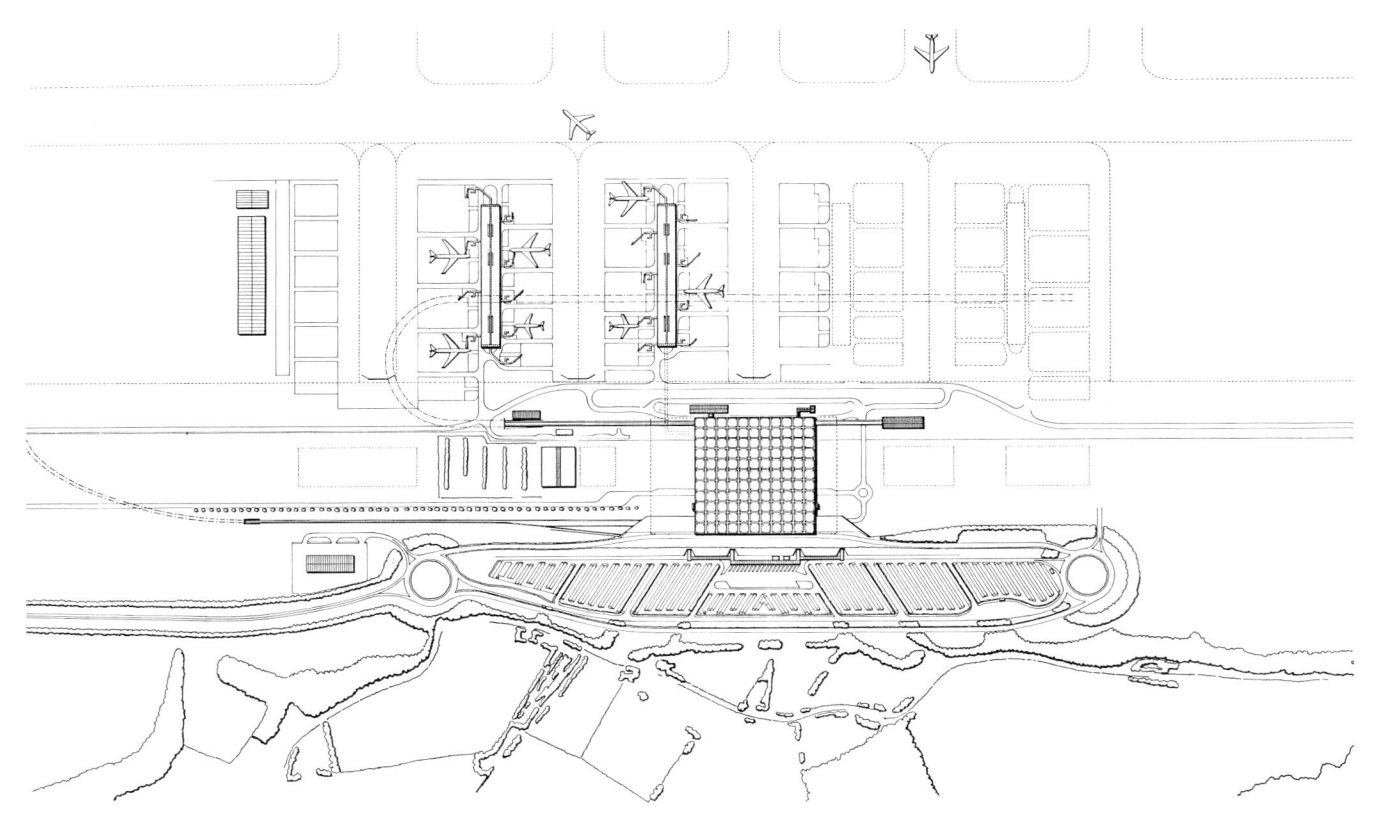

6.46 Stansted Airport Terminal area site plan (note terminal now extended and linked third pier added) (Courtesy of architects Foster & Partners).

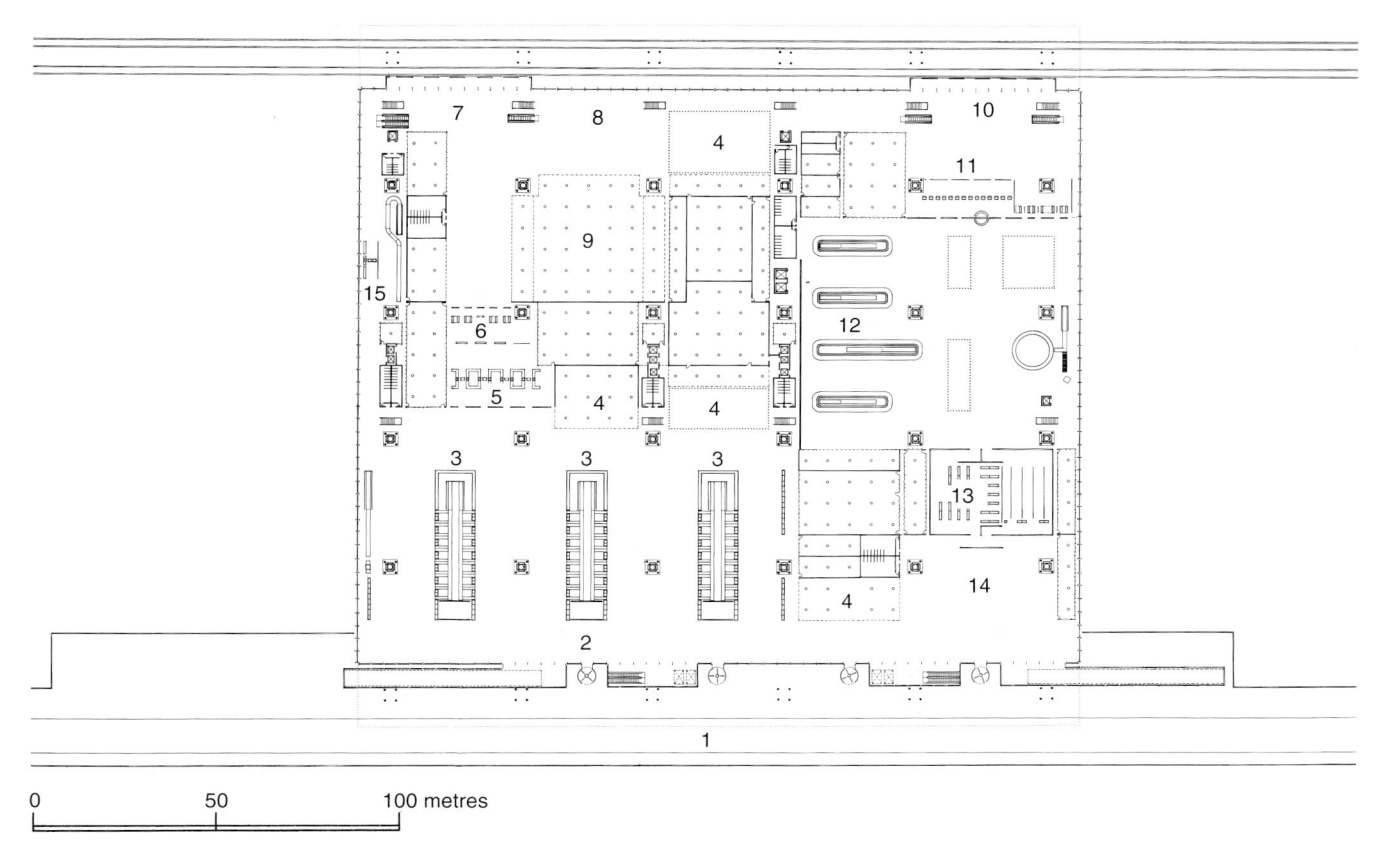

6.47 Terminal building concourse plan (Courtesy of architects Foster & Partners).

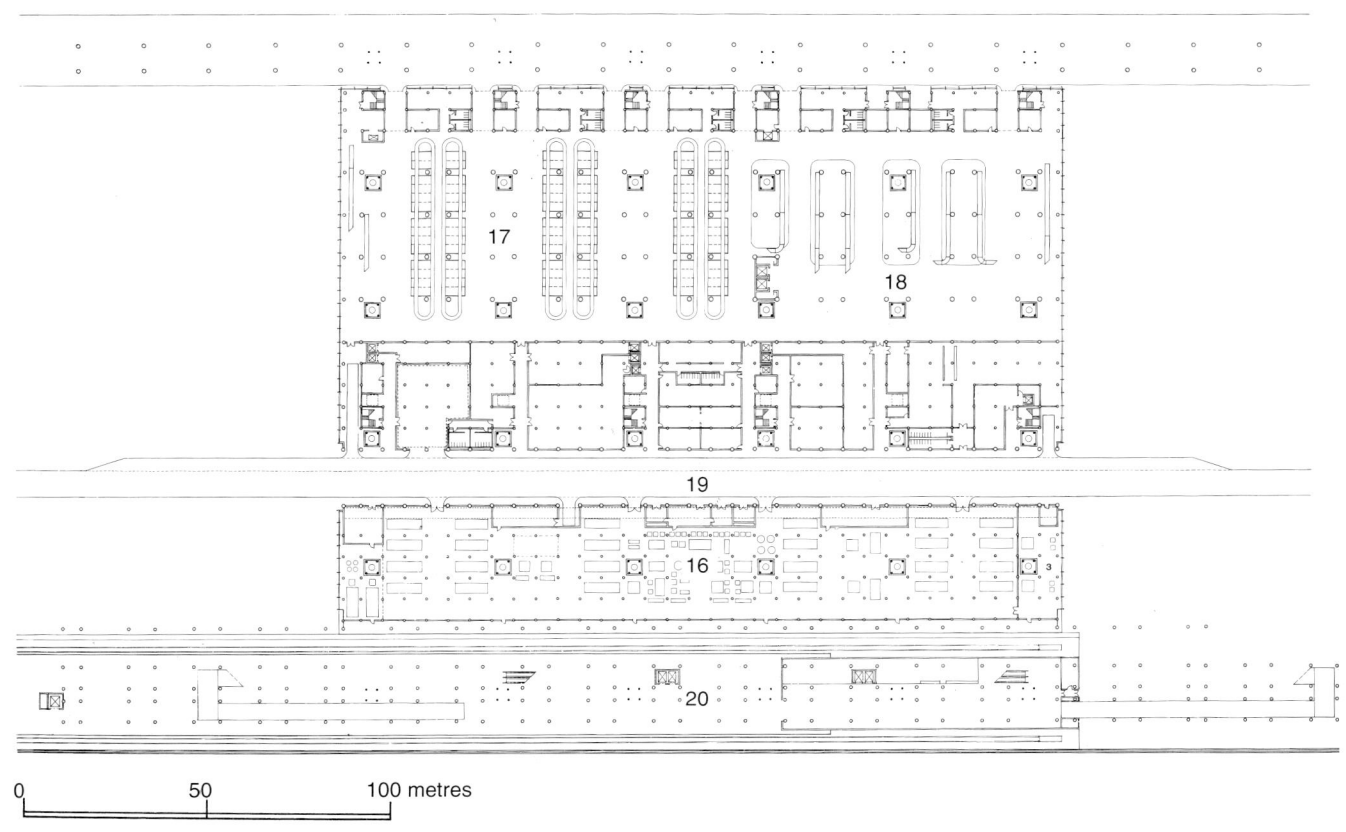

6.48 Terminal building undercroft plan (Courtesy of architects Foster & Partners).

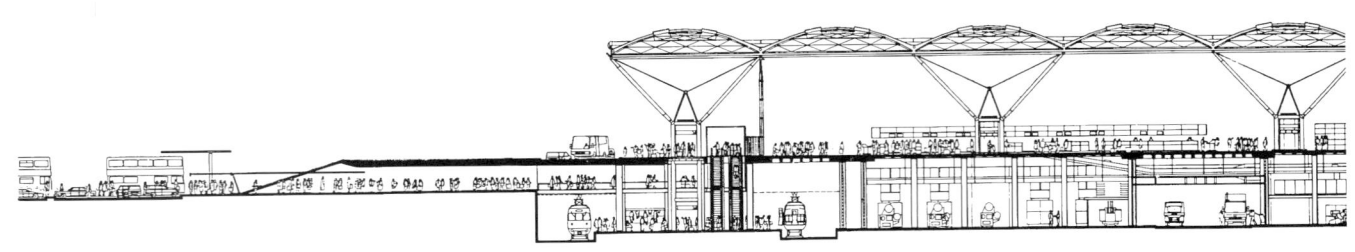

Transport terminals and modal interchanges

6.49 Section through rail station (left side), terminal and airside rapid transit (right side) (Courtesy of architects Foster & Partners).

6.50 View of station platform and escalator (from Stansted Express website).

Airport/railway interchange: contiguous

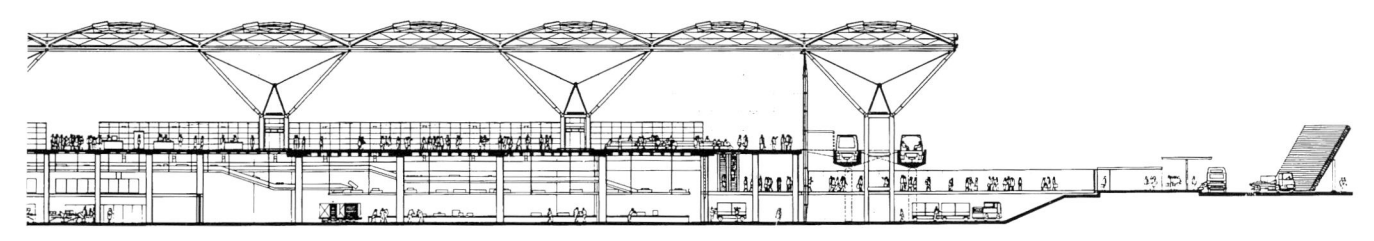

▲ 6.49 (Continued)

KEY TO TERMINAL PLANS (FIGURES 6.47–6.48)

1 Combined departures and arrivals access road
2 Departures hall
3 Check-in island
4 Shops and catering
5 Security control
6 Immigration control
7 Rapid transit departures
8 Departures lounge
9 Duty free shop
10 Rapid transit arrivals
11 Immigration control
12 Baggage reclaim
13 Customs
14 Arrivals hall
15 Domestic route
16 Plant
17 Departures baggage hall
18 Arrivals baggage hall
19 Service road
20 Railway station

TEAM

CLIENT: BAA plc
ARCHITECTS: Foster & Partners (original building) and Pascall & Watson Architects (extension 2001)

REFERENCE

GLA Planning Report on Proposed Development at Stansted Airport (2001). PDU/0267/01, 7 November.

6.2.4 Gatwick Airport, UK

Interchange – national railway station beside one terminal building and linked to other by people-mover, with bus and coach stations adjacent.

Between 1958 and 1977, Gatwick Airport's first (South) terminal was developed as a three-pier terminal handling predominantly charter traffic. In fact, the post-war terminal replaced the original terminal (described in Chapter 2) on a totally different site. The design capacity of the 120 m × 180 m central terminal building at the end of its growth cycle, including the enlarged Central Pier, was 11 million passengers per year.

A key feature of the South Terminal is the link with the adjacent railway station, which has 900 train services each day, including a dedicated service to London Victoria. More than 40 express coach services also serve the airport at a coach station beneath multi-storey car parks, both being connected to the South Terminal by moving walkways.

The northern pier was replaced in 1983 by the eight-stand circular satellite, which is linked to the central terminal by a rapid transit system. Many subsequent extensions have increased check-in areas and enhanced passenger lounges, catering and retail facilities, and the central pier has been converted to a vertically segregated system.

The second international terminal (North Terminal) supplements the original and expanded South Terminal, and in 1988 raised the planned capacity of the airport from 16 million to 25 million passengers per annum. The three-storey structure has a departures forecourt, check-in hall and departures lounge on the upper level. The middle level has the rapid transit station and a large shopping area on the landside and the inbound immigration facility on the airside. At apron level there are the baggage handling facilities, and customs and arrivals landside concourse and forecourt.

With a single-level 1988 airside concourse, reached by two spiral ramps from the departures lounge above, the subsequently required segregation of arrivals and departures created a problem. Screens and manned checkpoints ensure that arriving and departing streams do not mingle. A new pier added in 1991 has vertically segregated passenger levels, as has a further pier linked by a bridge over the taxiway serving the latter two piers.

The airport now handles 30 million passengers per year (ACI 2003 figure).

Modal split: of non-transfer passengers, 21% use the rail interchange (principally the Gatwick Express to Central London) and 9% the bus and coach interchange. Thus, a 30% public transport usage is being maintained but in a very different way from the counterpart airport at Heathrow. Sixty-eight per cent of passengers are UK-based leisure travellers, a much higher figure than at Heathrow (CAA 2002 figures).

TEAM, ORIGINAL TERMINALS (UP TO 1988)

CLIENT: BAA plc, formerly British Airports Authority
ARCHITECTS: YRM Architects and Planners

Transport terminals and modal interchanges

6.51 Gatwick Airport: aerial photo (courtesy of www.baa.com/photolibrary).

KEY TO AERIAL PHOTO (FIGURE 6.51)

B New pier and bridge added 2004
R Railway station
N North Terminal
S South Terminal

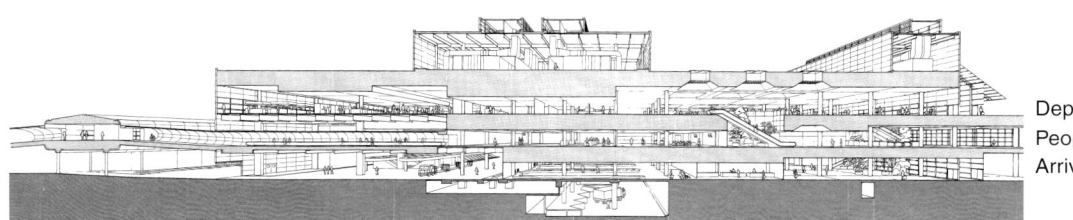

Departures
People-mover to South Terminal
Arrivals

6.52 North Terminal section (courtesy of YRM Architects and Planners).

Airport/railway interchange: contiguous **75**

6.2.5 New Hong Kong Airport at Chek Lap Kok

Interchange – high-speed airport railway terminus beside terminal building.

The airport now handles 27 million passengers per year (ACI 2003 figure).

In 2004 this airport won the accolade 'World's Best Airport' for the fourth year, awarded by Skytrax, the UK independent aviation research organisation.

The first phase of this terminal for an airport with an ultimate capacity of 80 million passengers per year is already a record-breaking project. It replaced the old terminal at Kai Tak, already reaching saturation in 1998 with an annual throughput of 24 million passengers served by a single runway. The new airport's first runway in Phase 1 will be followed by the second soon afterwards. The two runways are spaced 1525 metres apart, imposing severe constraints on the midfield terminal system.

The design of the building incorporates a Ground Transportation Centre with platforms for a dedicated high-speed rail link, the Airport Express, from the airport to urban Hong Kong on the landside and a sub-ground people-mover, which will ultimately link to a giant satellite on the airside.

The Airport Express offers a 23-minute service at 10-minute intervals, to Hong Kong Station and Kowloon stations of the Mass Transit Railway. Passengers can check in at these stations and from them free shuttle buses operate to hotels and other major public transport interchanges.

Overall public transport usage is always very high in Hong Kong. At the old Kai Tak Airport, passenger usage of taxis stood at 70% and now at the new airport bus usage stands at 70%.

Passengers from the direct ferry link to the PRC (People's Republic of China) transfer directly to the airside.

A vast commercial centre adjacent to the terminal area is termed SkyCity.

Airport design standards:

- International passengers per hour – Phase 1 capacity, 5500.
- Passengers per year – Phase 1 capacity, 35 million.
- Aircraft stands – Phase 1, 38 wide-body contact and 24 remote.

TEAM

CLIENT:	Hong Kong Airport Authority
ARCHITECTS:	Foster & Partners
STRUCTURAL ENGINEERS:	Arup and Mott MacDonald

▲ 6.53 New Hong Kong Airport at Chek Lap Kok: airport aerial master plan
(Source: website www.hongkongairport.com).

▲ 6.54 Terminal aerial view with ground transport interchange in foreground
(Source: website www.hongkongairport.com).

▲ 6.55 Terminal interior with people-mover (Source: website www.hongkongairport.com).

▲ 6.56 SkyCity (and ferry jetty) (Source: website www.hongkongairport.com).

6.2.6 Portland International Airport, USA

Interchange – urban light rail terminus beside terminal building.

This mid-range USA airport handles 12 million passengers per year and has been able to become one of only a few US airports with rail links thanks to a deal being struck between the transport authority and a developer. In return for development rights on land owned by the Port of Portland, the rail link was partially funded without federal funds and opened in 2001.

The light rail terminus of the Tri-met MAX Red Line, one of several urban rail lines, is directly beside the terminal building.

The City of Portland is pre-eminent among US cities for its light rail system, which has been planned and implemented in such a way that 95% of its passengers are car owners. Furthermore, the average citizen makes 61 trips per year by the light rail system – a high figure in the USA.

REFERENCE

Richards, B. (2001). *Future Transport in Cities*. Spon.

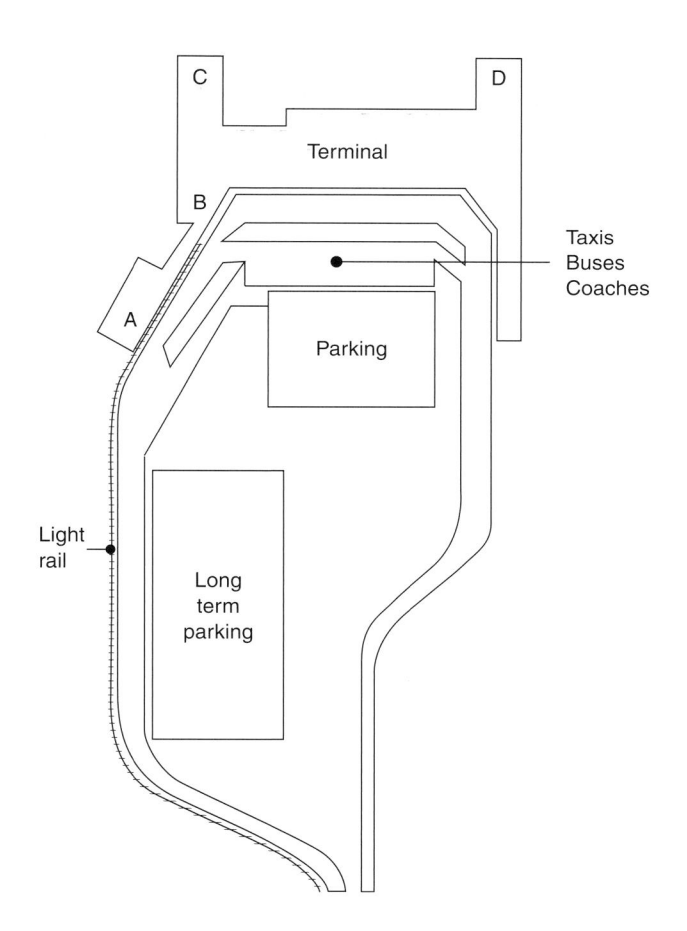

▶ 6.57 Portland International Airport: terminal diagram showing light rail.

Transport terminals and modal interchanges

6.58 Airport rail station (Courtesy of Tri-met).

6.59 Tri-met MAX Red Line at Portland Airport (Courtesy of Tri-met).

6.3 Airport/railway interchange: linked adjacent

6.3.1 Lyon St Exupéry Airport, France

Interchange – national and high-speed railway station linked to terminal buildings by moving walkway.

This airport has developed much (and been renamed in 2000 after the writer and aviator to commemorate the centenary of his birth, having previously been called Lyon Satolas) since opening in 1975. The airport is operated, like all provincial airports in France, by the local Chamber of Commerce as concessionaire in contract with the French government.

The original terminals were designed by Guillaume Gillet with a curved plan shape, the smaller as the Terminal National and the larger as the Terminal International. Then the TGV station was added in the early 1990s, connected to the terminals by moving walkways. The TGV station provides 18 high-speed trains per day, half serving Paris within 110 minutes.

The most recent project is the complete remodelling and extension of Terminal 2, formerly the domestic terminal, to take all traffic of Groupe Air France. This work was completed in 2003, and rounded off a programme to raise the airport's annual capacity from 4 million to 8 million passengers per year.

TEAM, TERMINAL 2

CLIENT:	Chambre de Commerce et d'Industrie de Lyon
ARCHITECTS:	Scott Brownrigg Ltd/CRB Architectes
CIVIL AND STRUCTURAL ENGINEERS:	Technip TPS
MECHANICAL AND ELECTRICAL ENGINEERS:	Technip TPS

The design team was commissioned to create a phased expansion programme, effectively turning the single-level domestic terminal into a vertically segregated four-channel terminal. The new need to distinguish traffic between Schengen Agreement states from other intra-European Union traffic has built the demand for four channels. The existing structure has been discontinued and the necessity for intricate phasing was imposed by the need to maintain the maximum passenger service throughout.

By virtue of its location, Lyon is a hub, and is served by a multiplicity of small airlines and aircraft, as well as pan-European and international flights. The link between the two terminals improves transfer times between international and domestic flights, and the target is 20-minute transfer times. The proximity of the TGV station not only improves surface links with the airport, but makes possible the integration of transport systems.

The planning of this transition has been particularly intricate. The original building was not easy to change. Plant originally located at ground level in the path of new essential baggage facilities has been relocated in a new basement constructed within the curtilage of the functioning terminal. The price of continuous operation is the long-term construction site.

Transport terminals and modal interchanges

▲ 6.60 Lyon St Exupéry Airport: aerial photomontage showing two terminals and TGV station (Courtesy of CRB Architectes and Scott Brownrigg Ltd).

▶ 6.61 Terminal 2 airside exterior (Courtesy of CRB Architectes and Scott Brownrigg Ltd).

Airport/railway interchange: linked adjacent **83**

Transport terminals and modal interchanges

6.62 Terminal 2 airside interior (Courtesy of CRB Architectes and Scott Brownrigg Ltd).

6.63 Terminal 2 landside interior (Courtesy of CRB Architectes and Scott Brownrigg Ltd).

84 Airport/railway interchange: linked adjacent

Transport terminals and modal interchanges

▶ 6.64 Exterior of TGV station, designed by Santiago Calatrava.

▲ 6.65 Interior of TGV station.

Airport/railway interchange: linked adjacent **85**

6.3.2 Birmingham Airport, UK

Interchange – national railway station and bus interchange linked to terminal building by people-mover.

The airport has two terminals, the main terminal (1984) and the Eurohub terminal for British Airways (1989).

(This review is based on a paper dated February 2003 given by Richard Heard, Managing Director, Birmingham International Airport Ltd.)

Birmingham International Airport is one of the UK's leading regional airports, currently serving in excess of 9 million passengers per annum. Passenger throughput is forecast to increase to 11 million over the next five years as the Airport serves an increasing proportion of its regional demand.

The Airport is located in the 'heart' of England and has a wide catchment across the Midlands. One of the key characteristics of the Airport's catchment (both inbound and outbound) is that it is distributed across a wide range of towns and centres around the full 360-degree circle from the Airport. This is unlike many capital city airports, where a high proportion of the catchment is from a single direction.

The wide spread of the catchment has made it particularly important for Birmingham International Airport to develop and support good facilities for all modes of surface access.

The need for good surface access is shared by the Airport's neighbour, the UK's National Exhibition Centre, which now attracts more than 5.5 million visitors per annum to an ever-increasing range of trade and retail exhibitions and shows. The co-location of these facilities has created the opportunity to develop a public transport interchange located adjacent to the Airport but also providing a new integrated transport hub for the region: Birmingham International Interchange.

Rail link

One of the unique advantages of Birmingham International Airport is its proximity to the UK's West Coast Mainline Railway. This is one of the UK's major 'trunk' railway routes, directly linking London to all the major cities on the west side of the UK. Major rail hubs along the route, of which Birmingham is one, link to the whole of the UK rail network.

Proximity to the rail line was one of the main drivers for the redevelopment of the Airport on its current site east of the airfield in 1984. At that time, the Passenger Terminal was linked to the rail station via an innovative Maglev (Magnetic Levitation) people-mover system. The Maglev system, unfortunately, became unserviceable in 1998 and was taken out of use. The system has now, however, been replaced with a new cable-drawn people-mover system to reinstate the direct link from the Airport to the rail station. The new people-mover system is supplied under a contract awarded to Doppelmayr, an Austrian company. The trains are able to carry 1500 passengers per hour on the 90-second journey.

The installation of a new people-mover system required the remodelling of the connection to the existing railway station. This, in turn, gave the opportunity to develop a complete new Multi-Modal Interchange, not only linking the Airport to the railway station, but also providing connections and facilities for buses, coaches, taxis and private car parking.

In addition to the two main users – the Airport and National Exhibition Centre – the new Interchange has been developed in partnership with the rail industry and regional planning and transportation authorities. Railtrack (now Network Rail) made the site for the new facility available. Virgin West Coast Trains – the operator of the existing station – has made possible the temporary changes to the station required to enable construction. The train operator has also assisted in drawing up and agreeing new operating arrangements for the combined existing station and new Interchange. The local planning and transportation authorities provided substantial grant aid through their Local Transport Plan support package and a contribution came from the European Commission's Trans European Networks (TENS) fund.

The Interchange facility

The new Interchange provides approximately 2000 m^2 of internal floor space spread across two floors and straddling the station forecourt roads. On the Ground Floor, new bussing lounges are provided to each side of the forecourt roads. There is also a separate exit to a new taxi rank and routes in from car parks and pedestrian walkways to the station. Both of the bussing lounges are linked to the First Floor with escalators, steps and lifts – all fully complying with UK and EC disabled access requirements. The First Floor provides the main concourse, giving direct access to the railway station, National Exhibition Centre and, via the new people-mover system, to the Airport Passenger Terminals. The First Floor also has the potential to provide airline check-in and ticket desk facilities, along with automated check-in machines.

Modal split

The declared aim is 20% usage of public transport, and a weighting is applied to percentage use by passengers, employees and visitors of 0.65, 0.30 and 0.05 respectively to calculate achievement of that aim. Closure of the Maglev caused public transport usage to fall to 11% in 1999, but it is now on the increase again.

TEAM, INTERCHANGE

CLIENT: Birmingham International Airport Ltd
ARCHITECTS: CPMG Architects

Transport terminals and modal interchanges

▲ 6.66 Birmingham International Airport: aerial photo, showing airport and NEC (Courtesy of Birmingham International Airport Ltd).

Airport/railway interchange: linked adjacent

Transport terminals and modal interchanges

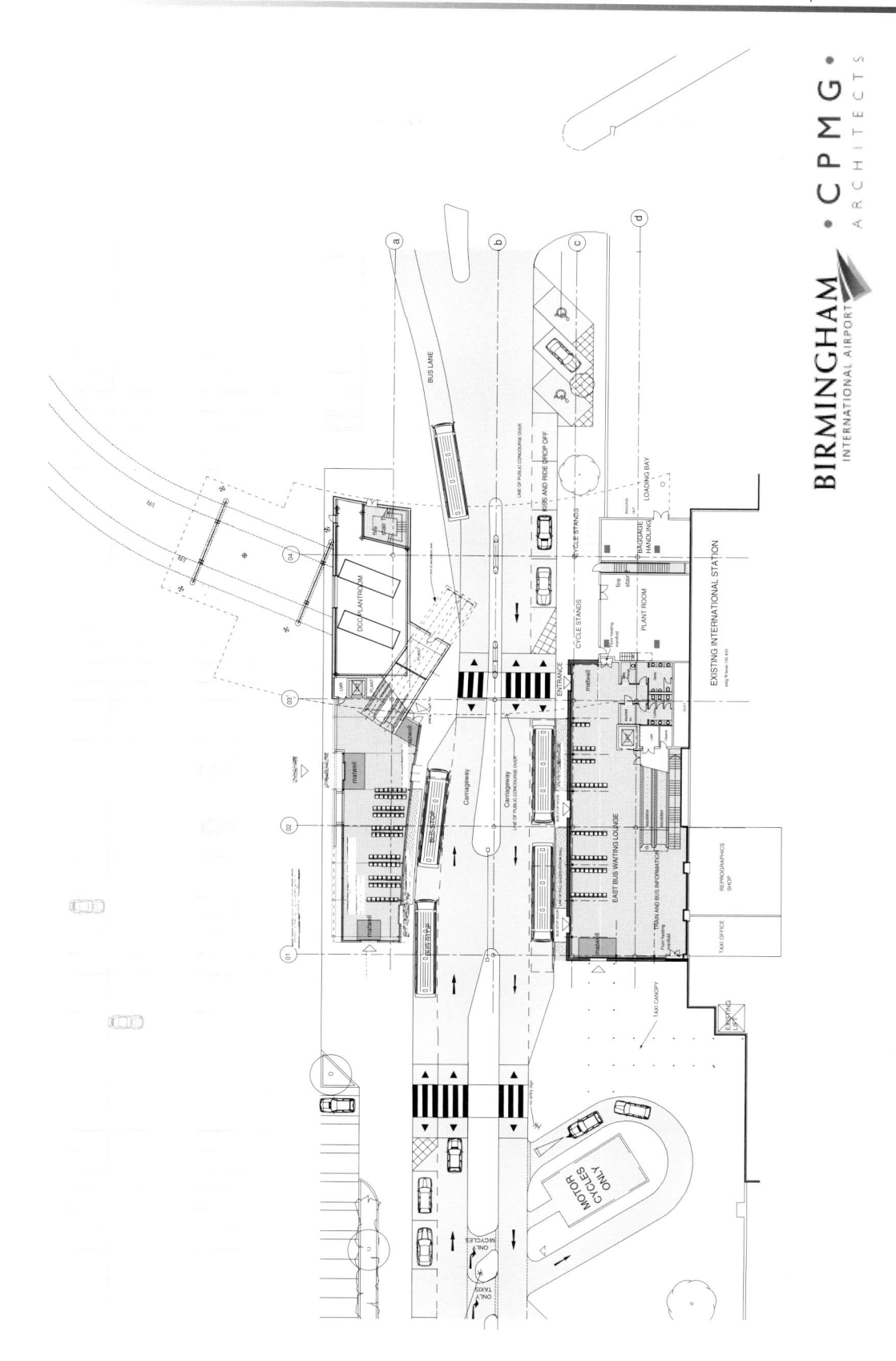

▲ 6.67 Interchange Ground Floor plan (Courtesy of Birmingham International Airport Ltd).

Airport/railway interchange: linked adjacent **89**

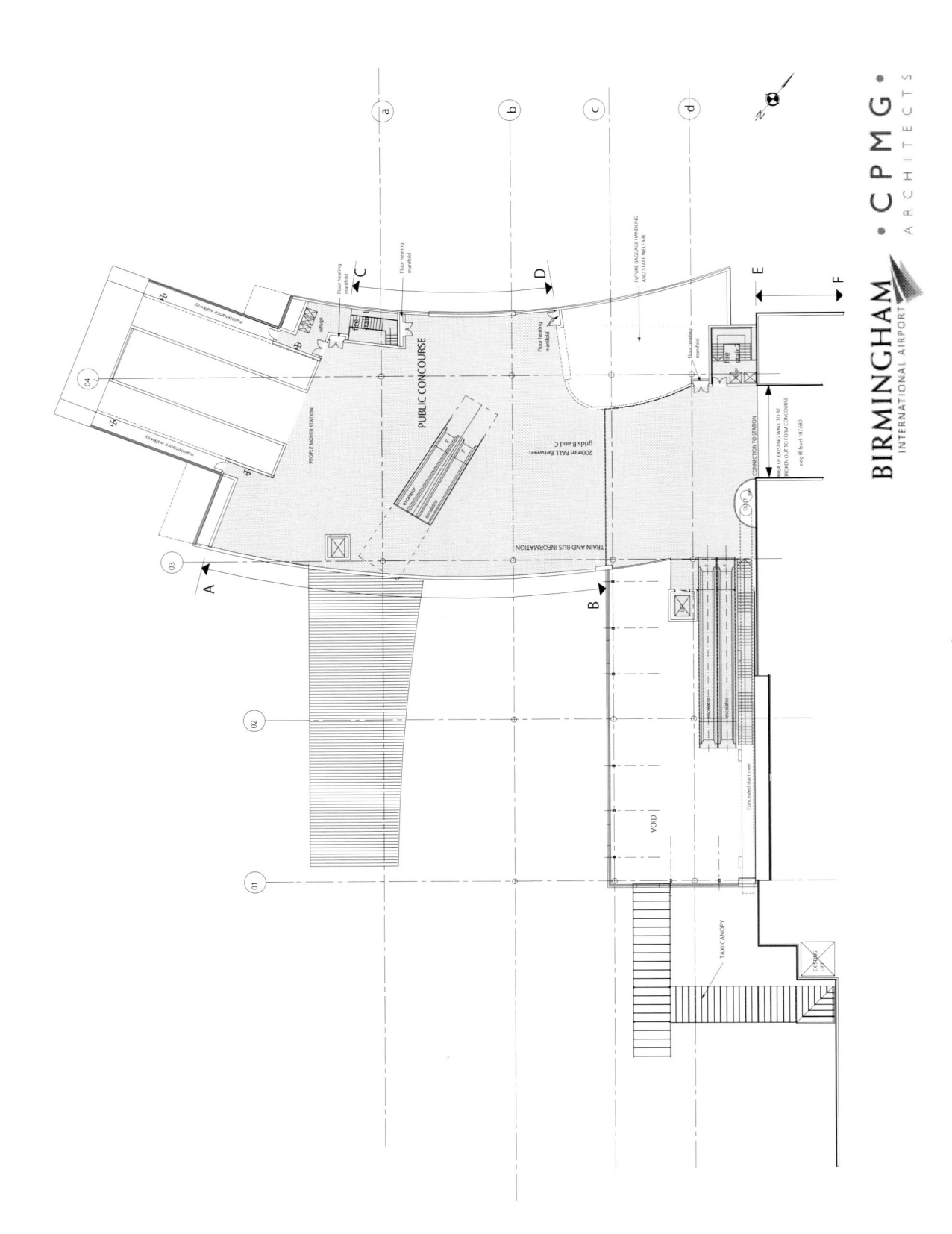

Transport terminals and modal interchanges

▲ 6.68 Interchange First Floor plan (Courtesy of Birmingham International Airport Ltd).

90 Airport/railway interchange: linked adjacent

Transport terminals and modal interchanges

▶ 6.69 People-mover at Interchange (Courtesy of Birmingham International Airport Ltd).

▶ 6.70 People-mover in transit (Courtesy of Birmingham International Airport Ltd).

▶ 6.71 Interchange building: people-mover on the left, main rail station on the right (Courtesy of Birmingham International Airport Ltd).

REFERENCE

Heard, R. (2003). Presentation at Passenger Terminal Expo, Hamburg.

Airport/railway interchange: linked adjacent

6.3.3 Manchester Airport, UK

Interchange – urban railway terminus and bus station connected to three terminal buildings by long corridors and moving walkways.

The whole airport now handles 20 million passengers per year (ACI 2003 figure).

From the new Ground Transport Interchange (GTI), which incorporates the rail station opened in 1993, six trains per hour offer a 15-minute service to Manchester city centre and bus routes radiate throughout the region. The airport's target is that 40% of passengers will travel to and from the airport by public transport by 2015. Provision is made in the lowest level of the GTI for extension of the rail line and addition of the metropolitan light rail system, Metrolink. This integrated facility is the result of a partnership between Manchester Airport plc, the Greater Manchester Transport Executive, the Strategic Rail Authority and the rail, bus and coach operators.

A ground-level concourse serves the bus and coach station and the rail station below, and Skylink, an upper concourse, connects with links to the three terminals.

▲ 6.72 Manchester Airport: aerial photo.

Transport terminals and modal interchanges

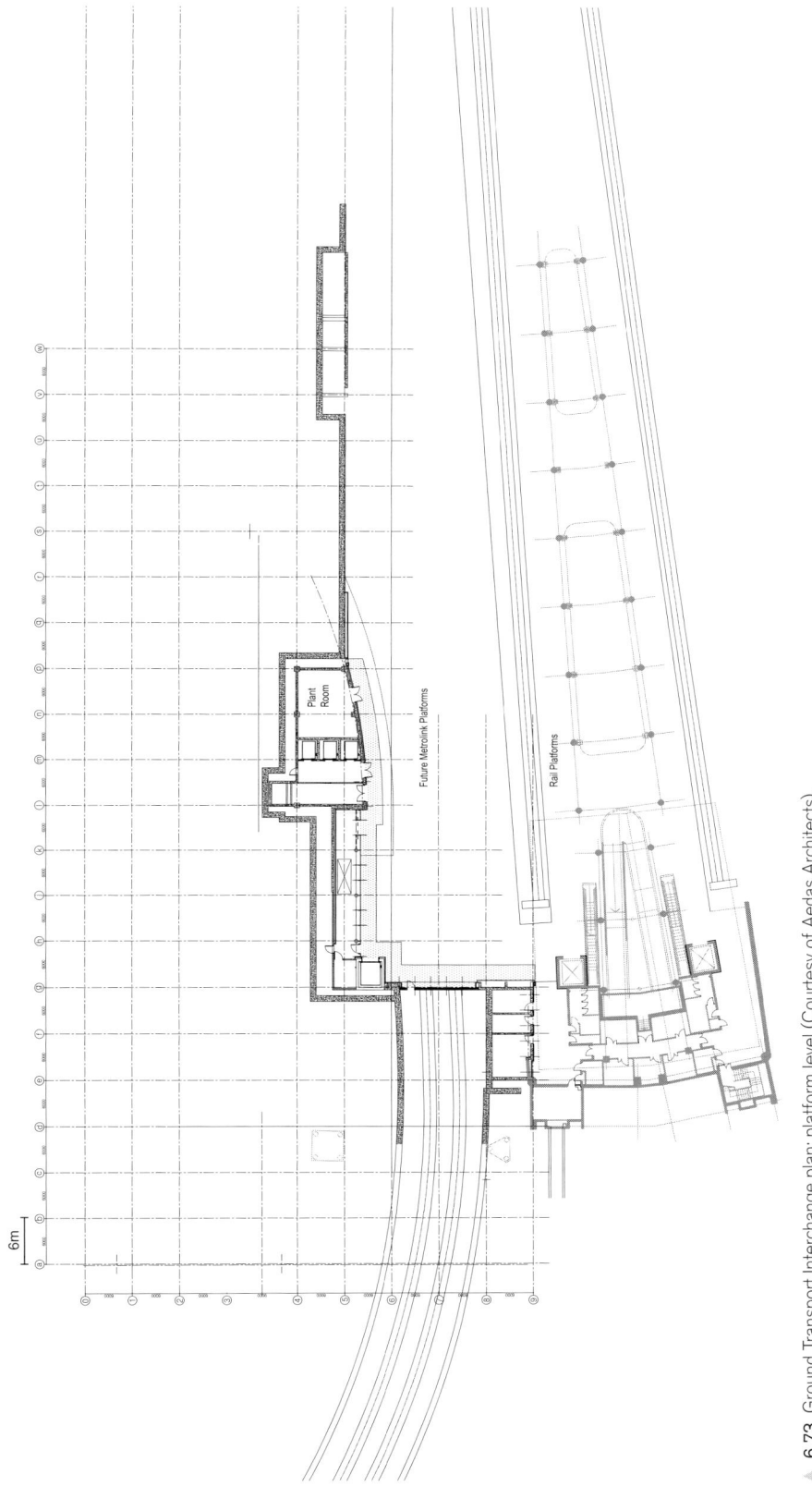

▲ 6.73 Ground Transport Interchange plan: platform level (Courtesy of Aedas Architects).

Airport/railway interchange: linked adjacent **93**

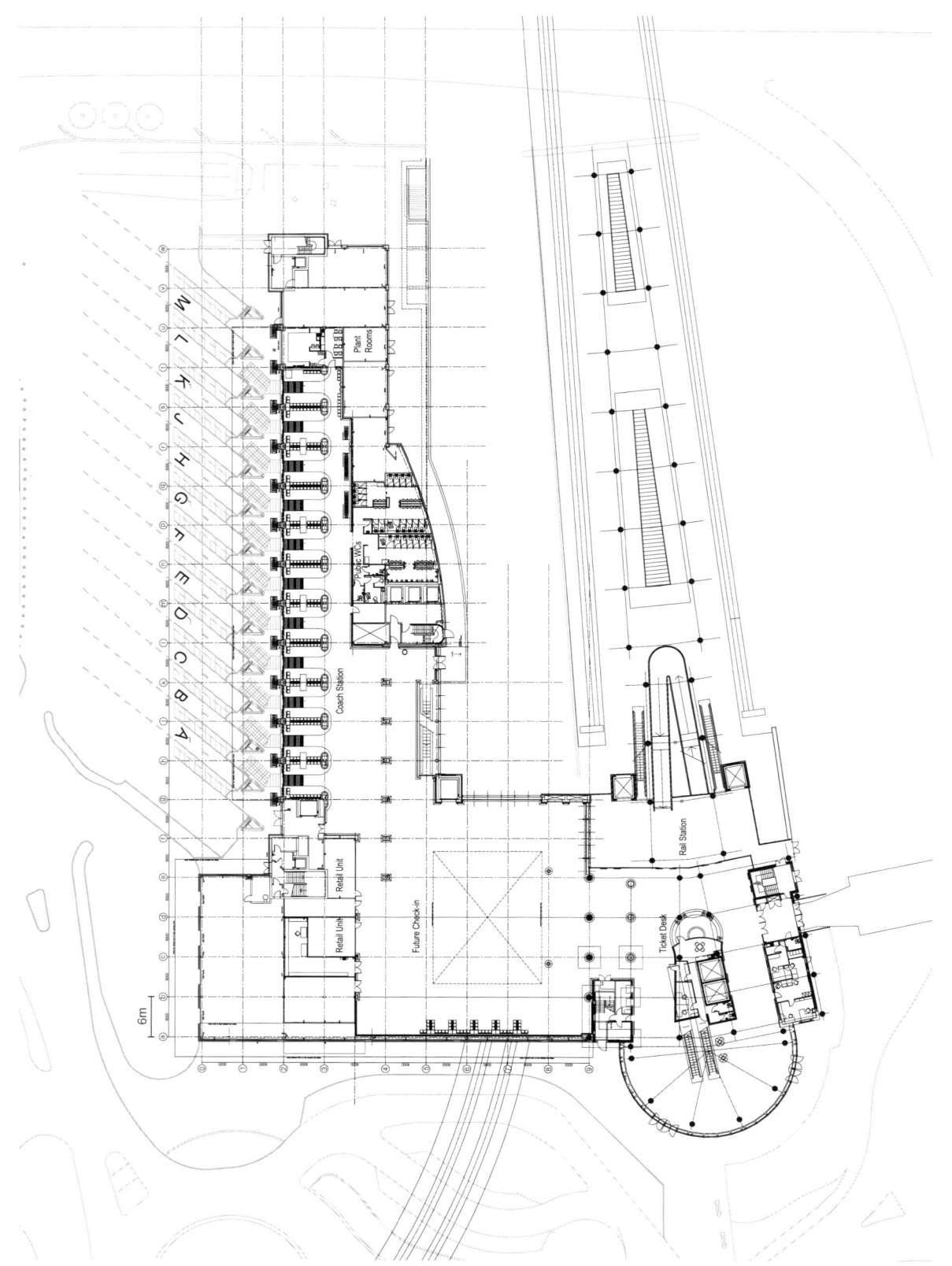

Transport terminals and modal interchanges

▲ 6.74 Ground Transport Interchange plan: concourse (Courtesy of Aedas Architects).

94 Airport/railway interchange: linked adjacent

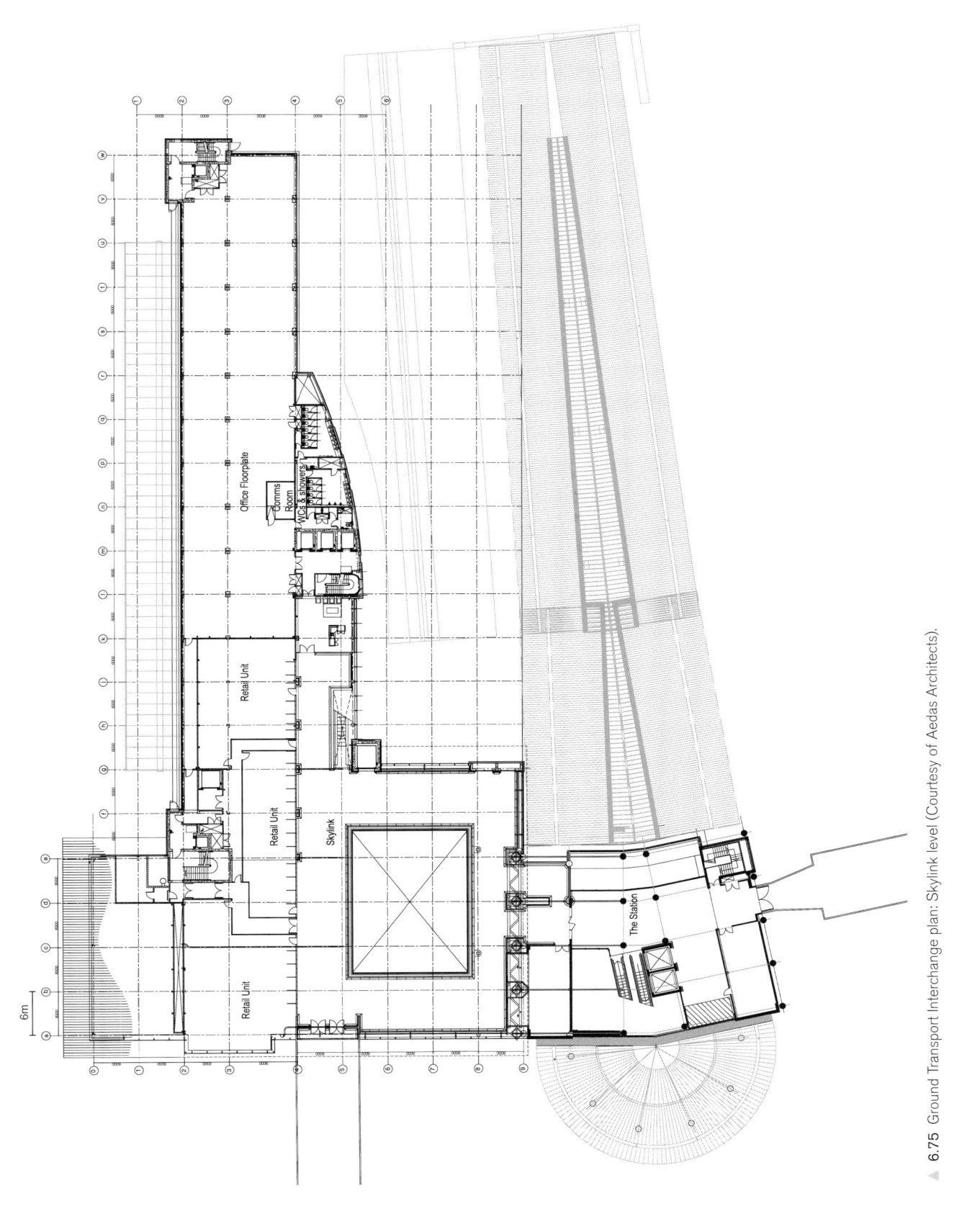

6.75 Ground Transport Interchange plan: Skylink level (Courtesy of Aedas Architects).

Transport terminals and modal interchanges

▶ 6.76 Ground Transport Interchange: exterior view (Courtesy of MA plc).

▶ 6.77 Ground Transport Interchange: rail station view (Courtesy of MA plc).

TEAM, GROUND TRANSPORT INTERCHANGE

CLIENT: Manchester Airport Developments Ltd
ARCHITECTS: Aedas Architects

Airport/railway interchange: linked adjacent

Transport terminals and modal interchanges

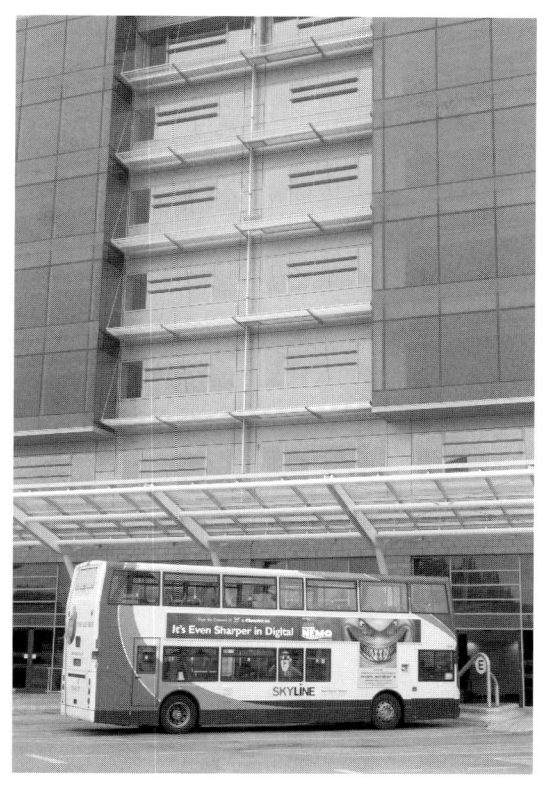

6.78 Ground Transport Interchange: bus station view (Courtesy of MA plc).

6.79 Ground Transport Interchange: concourse view (Courtesy of MA plc).

6.80 Ground Transport Interchange: upper concourse view (Courtesy of Chris Chalk of Mott MacDonald).

6.81 Ground Transport Interchange: bus bays (Courtesy of Chris Chalk of Mott MacDonald).

Airport/railway interchange: linked adjacent **97**

Manchester Airport's first terminal, together with the domestic pier known as Terminal 3 opened in May 1989, has a capacity of 12 million passengers per year. The new second terminal, the first phase of which opened in 1993, is Manchester's answer to a growth strategy of additional long-haul routes, traffic which otherwise might be routed via one of the overcrowded airports of the south-east of England. The capacity of the second terminal is to grow to 12 million passengers per year in two phases. The site constraints have determined that the first phase shall comprise most of a central terminal and a long single-sided pier; the second phase plus a remote island two-sided pier will comprise the remainder.

The building offers segregation of arriving and departing passengers: from the elevated departures forecourt to check-in, outbound controls and then the airside concourse, passengers have a level route. A series of stand access towers on the airside face provide stairs and lifts down to the loading bridges. Arriving passengers have a level route into an arrivals corridor which leads to the inbound central immigration control and thence down to baggage reclaim, customs and the landside concourse.

When Phase 2 is constructed, a series of escalators and lifts will link the passenger levels of the central building with a tunnel to the remote pier.

Terminal 2 design standards:

- International passengers per hour – 1850 each way (at 6 million passengers per year) and 2700 each way (at 12 million passengers per year).
- Transit/transfer passengers – 15%.
- Passengers per year – 6 million (Phase 1) plus 6 million (Phase 2).
- Aircraft stands – eight B747-400 (or B757/B757 or B767/B737 on each) at Phase 1, and 14 B747-400 plus five MD11, including the remote pier, at Phase 2.

TEAM, TERMINAL 2

CLIENT:	Manchester Airport plc
ARCHITECTS:	Scott Brownrigg
STRUCTURAL ENGINEER:	Scott Wilson
SERVICES ENGINEER:	Oscar Faber & Partners (now Faber Maunsell)

▶ 6.82 Terminal 2 landside exterior view (Courtesy of architects Scott Brownrigg Ltd).

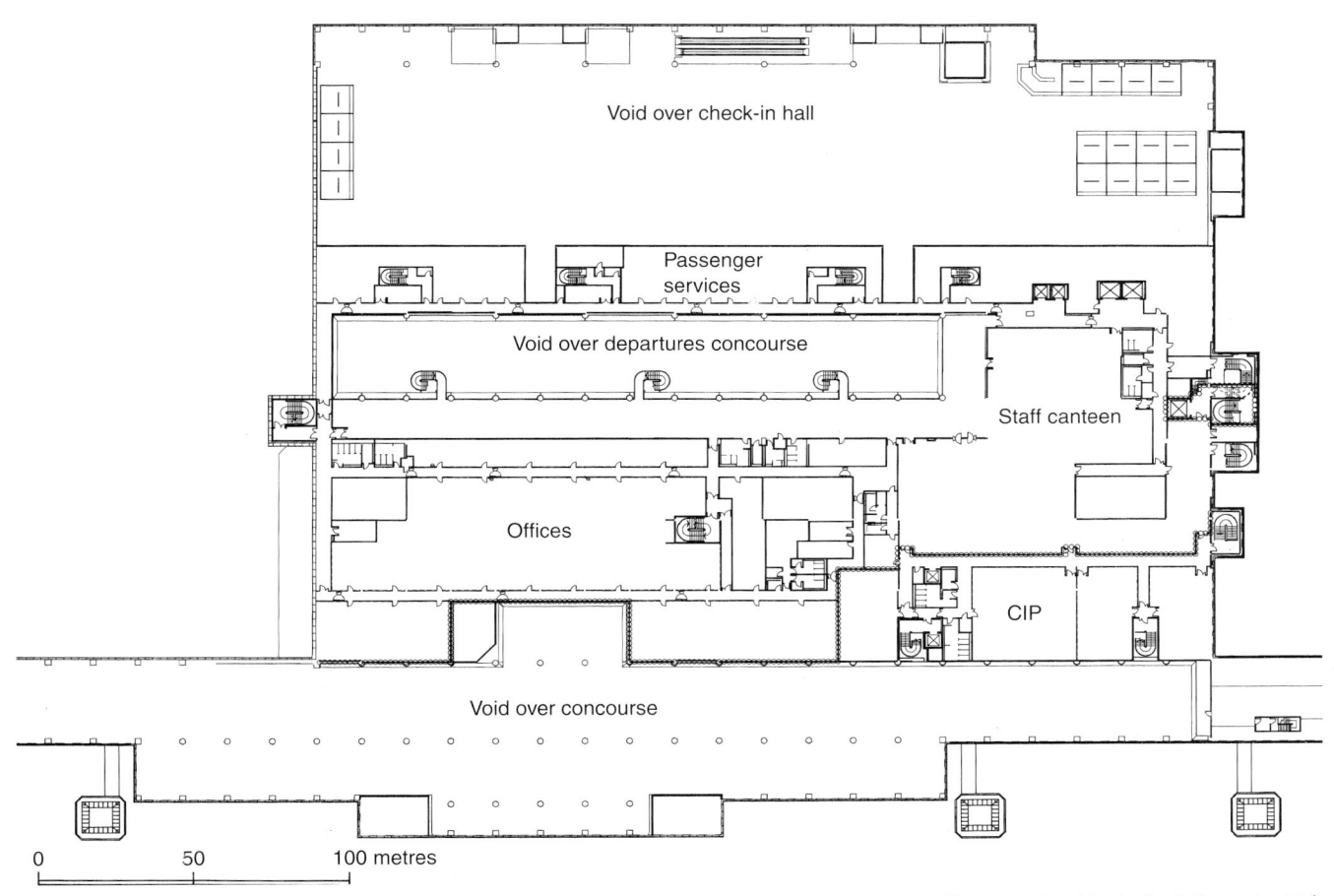

Transport terminals and modal interchanges

▲ 6.83 Departures mezzanine level plan (Courtesy of architects Scott Brownrigg Ltd).

▶ 6.84 Terminal 2 landside interior view (Courtesy of architects Scott Brownrigg Ltd).

Airport/railway interchange: linked adjacent **99**

Transport terminals and modal interchanges

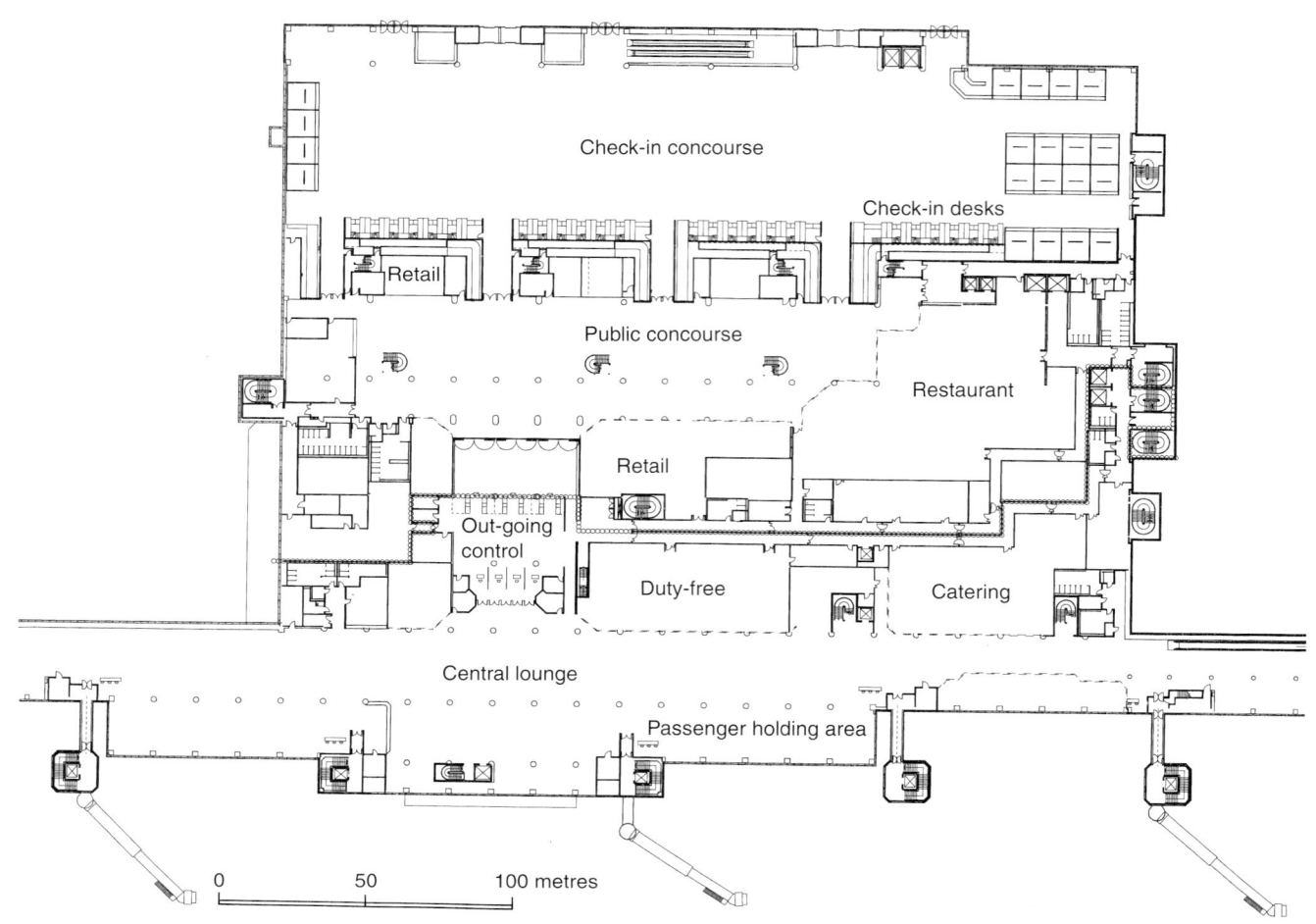

▲ 6.85 Departures level plan (Courtesy of architects Scott Brownrigg Ltd).

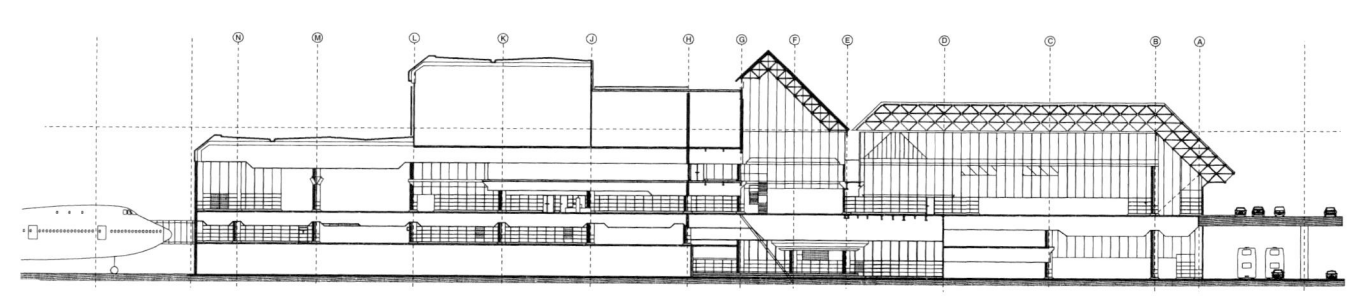

▲ 6.86 Terminal 2 section (Courtesy of architects Scott Brownrigg Ltd).

100 Airport/railway interchange: linked adjacent

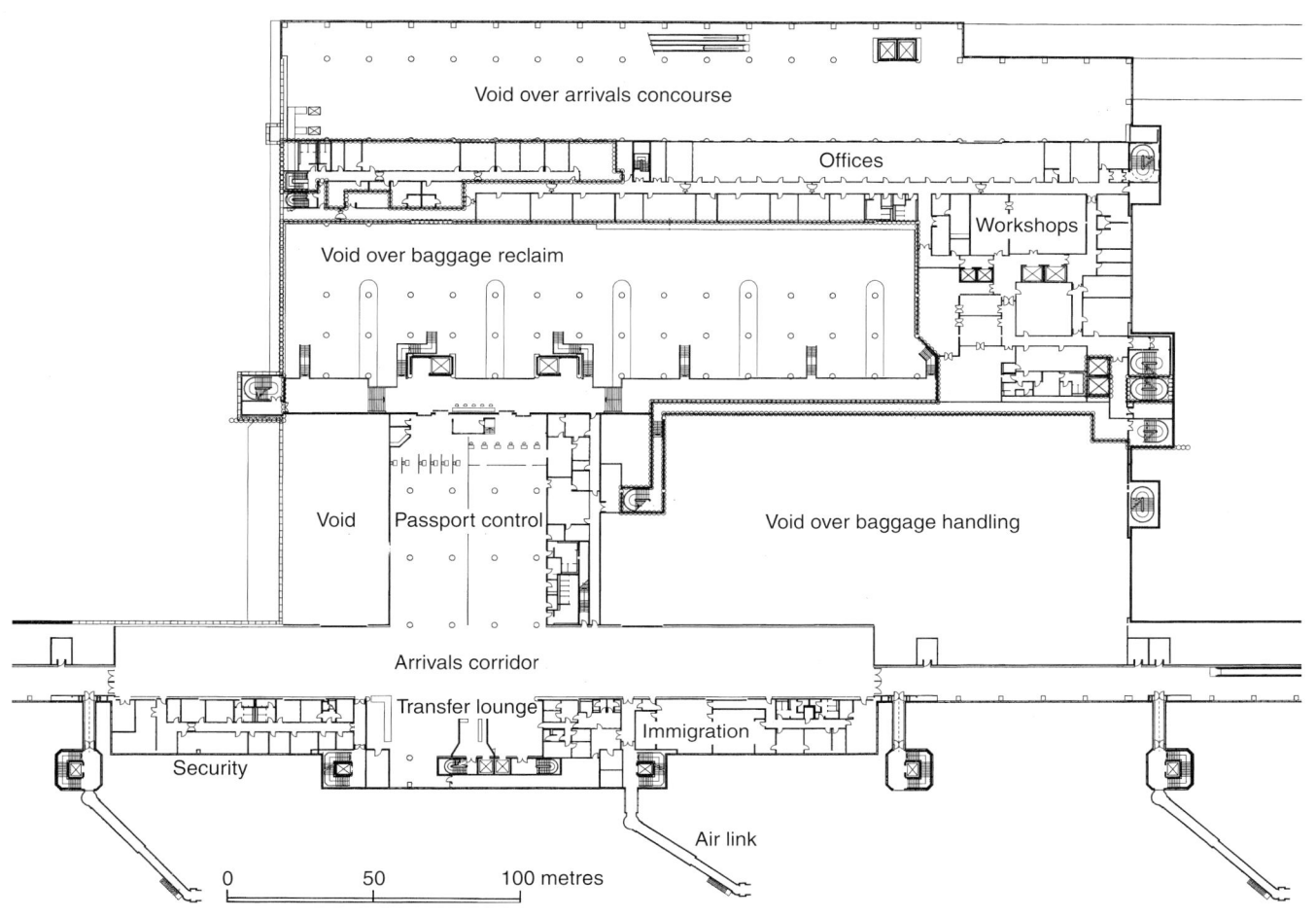

Transport terminals and modal interchanges

6.87 Arrivals mezzanine level plan (Courtesy of architects Scott Brownrigg Ltd).

Airport/railway interchange: linked adjacent **101**

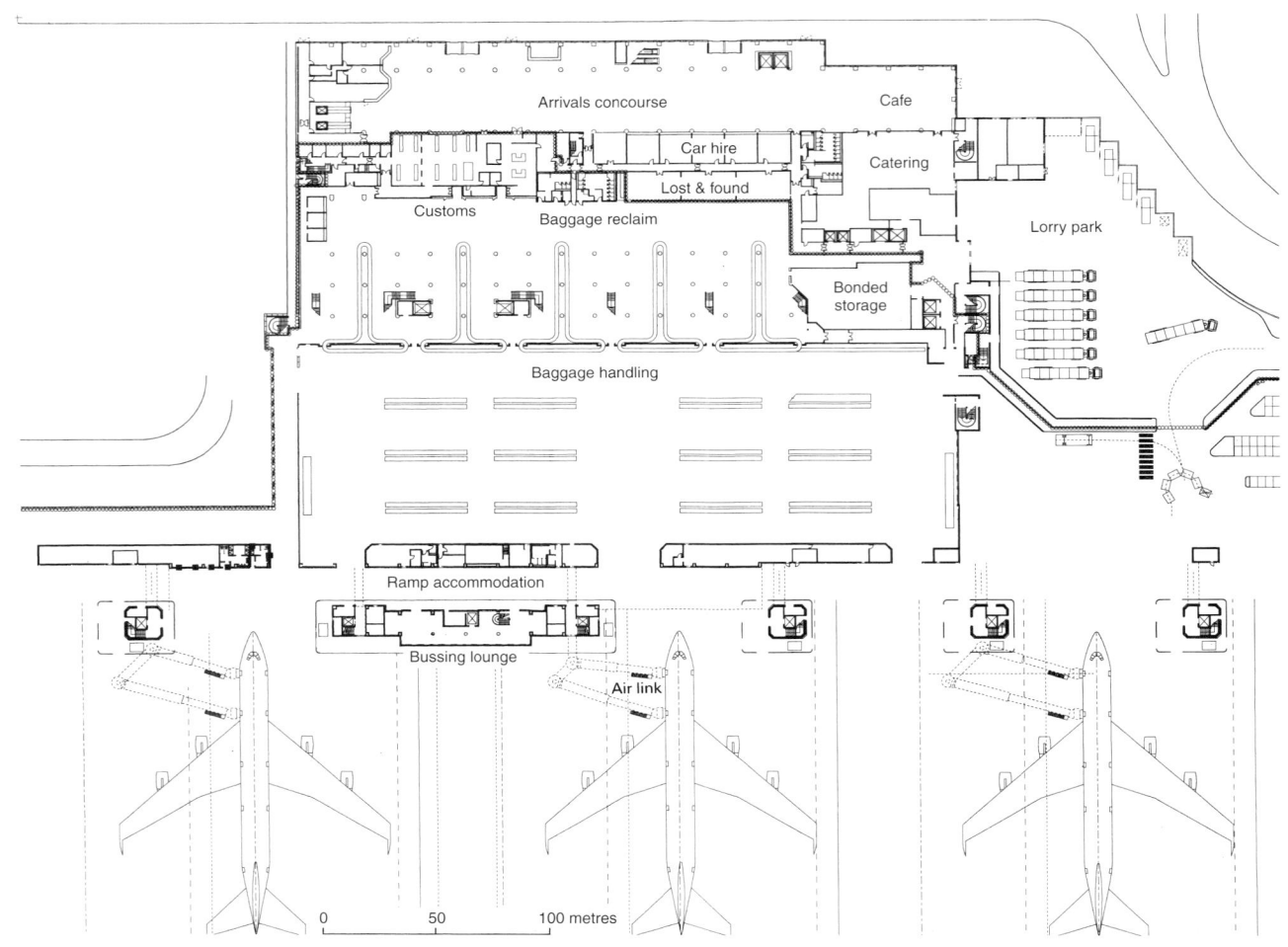

Transport terminals and modal interchanges

▲ 6.88 Arrivals level plan (Courtesy of architects Scott Brownrigg Ltd).

REFERENCES

Airports International (2002). Ground Transport Interchange. November/December.
Church, J., Aplin, D. et al. (1993). Interiors study, Terminal 2. *Architects Journal*, 19 May.
Longworth, R. (2003). Presentation at Future Terminal Conference, London.
Pearson, S. (2003). Presentation at Passenger Terminal Expo, Hamburg.

102 Airport/railway interchange: linked adjacent

6.3.4 Heathrow Airport Central Terminal Area, UK

Interchange – metro station and large bus station connected to three terminal buildings by long corridors and moving walkways. Also dedicated rail link to central London (Heathrow Express).

The whole airport now handles 63.4 million passengers per year (ACI 2003 figure).

The three central area terminals were developed originally in the 1950s and 1960s in the centre of a star-shaped runway pattern and connected to the north by a public road tunnel and to the south by a private cargo tunnel. A separate fourth terminal (1986) and fifth terminal (due to open 2008) have separate landside road approaches and public transport links.

The London Underground Piccadilly Line was extended to Heathrow in 1977/8, leading to the quotation 'the train now arriving at Heathrow is 41 years late', an allusion to the opening of the original Gatwick 41 years earlier (see Chapter 2). Until the extension to Terminal 4 in 1986, there was a problem of lack of space to store trains in the terminus in the central area. Both the Piccadilly Line and the Heathrow Express tunnels now run through the central area to Terminal 4.

Thirteen per cent of passengers use the Piccadilly Line (CAA 2002 figure) and 9% of passengers use the Heathrow Express, which provides a 15-minute service to Paddington every 15 minutes (CAA 2002 figure). A further 13% of passengers use buses and coaches to and from the busiest bus station in the country at the heart of the central area.

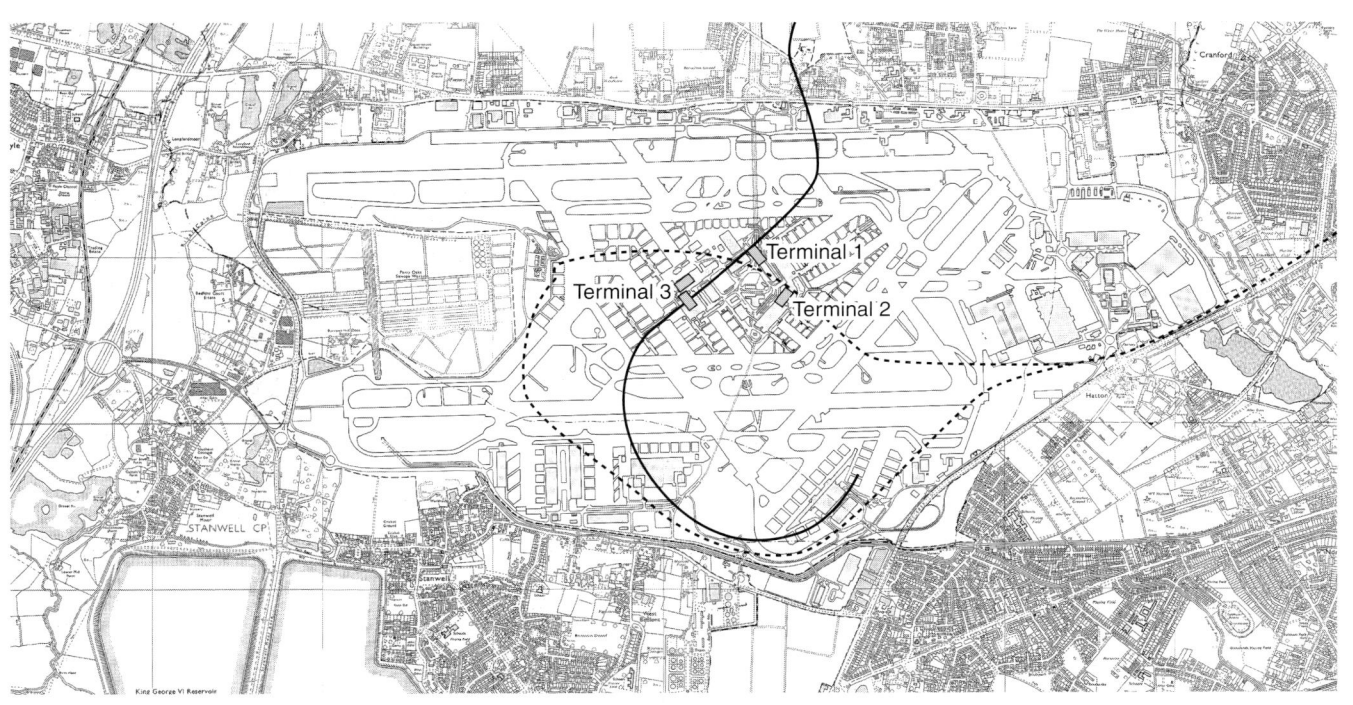

6.89 Heathrow Airport Central Terminal Area with Heathrow Express (continuous line) and Piccadilly Line (dotted line) approach routes (Courtesy of Heathrow Airport Ltd).

Transport terminals and modal interchanges

▶ 6.90 Heathrow Express image (Courtesy of Chris Chalk of Mott MacDonald).

▶ 6.91 Bus station from above.

▶ 6.92 Bus station entrance with lift from Underground station.

104 Airport/railway interchange: linked adjacent

Transport terminals and modal interchanges

6.93 Long-distance bus stands.

6.94 Local bus stands.

REFERENCE

BAA Heathrow (2003). Shaping the vision: a surface access strategy for Heathrow. Progress Report, October.

Airport/railway interchange: linked adjacent **105**

6.3.5 Southampton Airport, UK

Interchange – national railway station and bus stands adjacent to terminal building.

Apart from the fact that Southampton (Eastleigh) – now Southampton International Airport – is the smallest BAA airport, it has three special features, all of which have influenced the design of the new single-level terminal. Capacity is 1.5 million passengers per year and 350 passengers per hour, served by 14 aircraft stands.

Firstly, its location as a natural bridgehead to the Channel Islands: this has acted as the mainstay of the airport and generates 50% of the traffic, hence the high demand for duty-free shopping.

Secondly, its location as a regional airport for the southern part of Hampshire and a catchment area of nearly 2 million people, offering many mainland, Irish or other European destinations, as well as the Channel Islands. This range of traffic demands five different sets of border control procedures in one simple building. These are: Domestic, Belfast (with special branch surveillance), Common travel area (Channel Islands, Isle of Man and Ireland), EC and European/International non-EC.

Thirdly, and not least, its location adjacent to Southampton Parkway railway station, served by fast trains to London Waterloo. Although only 11% of passengers use the rail link at present, this must be a growth area, and there are plans to add more platforms to the station.

TEAM

CLIENT:	BAA plc
ARCHITECTS:	The Manser Practice
STRUCTURAL ENGINEER:	Peter Brett Associates

REFERENCE

Blow, C.J. et al. (1995). Building study. *Architects Journal*, 30 March.

▶ 6.95 Southampton Parkway railway station with bus and car parking in front of airport.

Transport terminals and modal interchanges

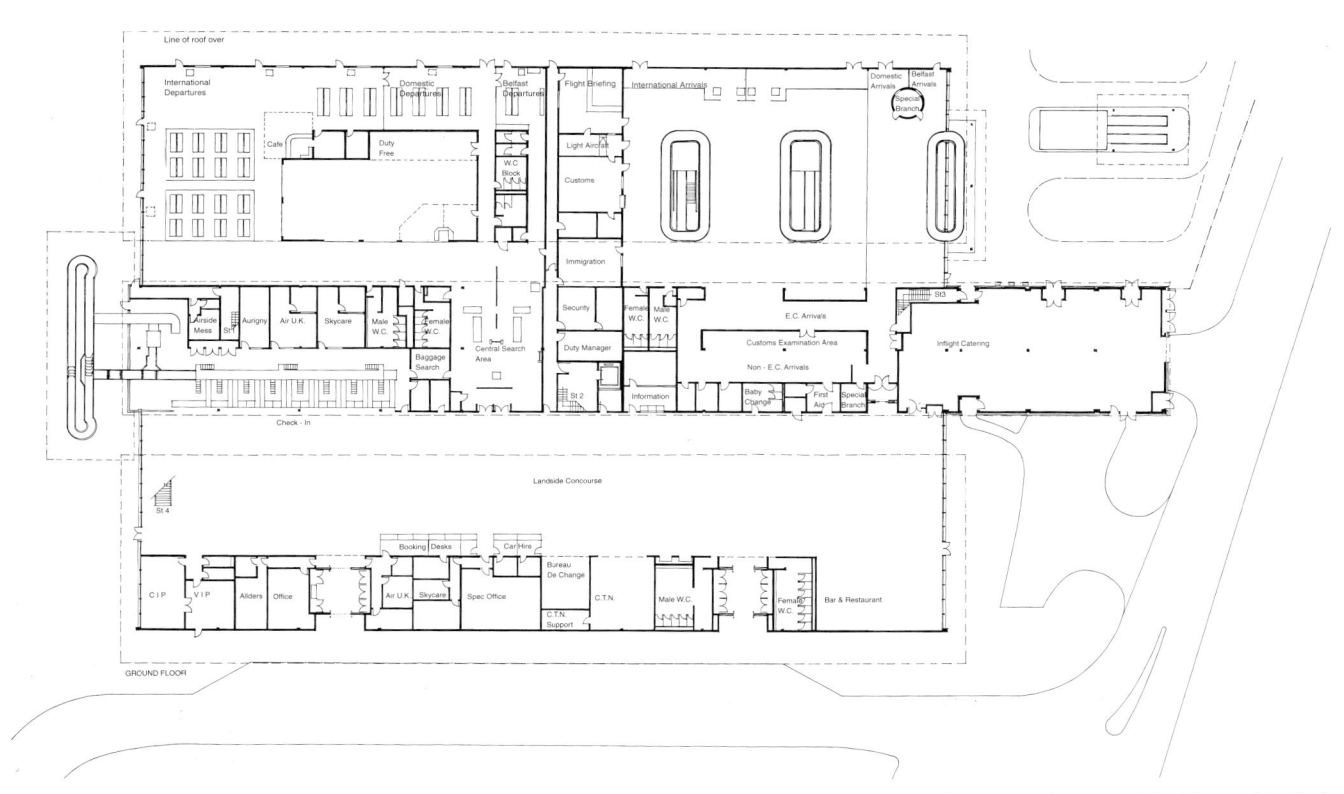

6.96 Airport terminal ground floor plan (Courtesy of architects The Manser Practice).

Transport terminals and modal interchanges

6.3.6 Atlanta Hartsfield-Jackson Airport, USA

Interchange – metro railway terminus adjacent to central terminal building.

Now the busiest airport in the world with 79 million passengers per year (ACI 2003 figure), this was almost the first in the USA in 1980 to have a potential metropolitan rail link 'built in'. The rail station was actually opened in 1988. The MARTA terminus outside the central terminal offers a 15-minute ride to Downtown Atlanta every 8 minutes, and achieves 9% usage by originating and departing passengers, bearing in mind that the airport serves 70% transfers.

Atlanta is located as an ideal interchange point in the air transport system of the south-eastern United States. When planned as a major hub, Atlanta was second only to Chicago in the number of domestic passengers handled, and has now overtaken Chicago.

All originating passengers arrive at the kerbside of one of two twin landside terminals. From here, passengers are security-cleared centrally and descend into an underground mall. They travel by moving pavement or rapid transit system to five concourses, each 660 m (2200 ft) long and 300 m (1000 ft) apart.

TEAM, ORIGINAL TERMINAL

ARCHITECTS: Stevens & Wilkinson/Smith, Hinchman & Grylls/Minority Airport Architects and Planners

▲ 6.97 View of rail station (Courtesy of atlanta-airport.com).

▲ 6.98 Atlanta Hartsfield-Jackson Airport: visualisation of airport terminal area (Courtesy of atlanta-airport.com).

Airport/railway interchange: linked adjacent

6.3.7 Ronald Reagan Washington National Airport, USA

Interchange – metro railway station bridge-linked to two out of three terminal buildings.

The airport now handles 14 million passengers per year (ACI 2002 figure).

The original terminal was built in 1941 and subsequently extended before the two new linked terminals were built in 1999 to design by Cesar Pelli and Leo Daly. The continuous concourse level of Terminals B and C, connected to the Metrorail station, is 500 metres long and acts like a street, with the check-in level above and the baggage reclaim below. The three new piers of Terminals B and C have added over 30 terminal-served stands.

The original 1977 Metrorail subway station at Washington National was built remote from the then terminal (A), in spite of the fact that the US Department of Transportation ruled both the Federal Aviation Administration and the Urban Mass Transit Administration. However, when new terminals (B and C) were built they were put next to the Metrorail station and by 2004 the modal split had risen above 10% by Metrorail.

TEAM, TERMINALS B AND C

CLIENT: Metropolitan Washington Airport Authority
ARCHITECTS: Cesar Pelli and Leo Daly

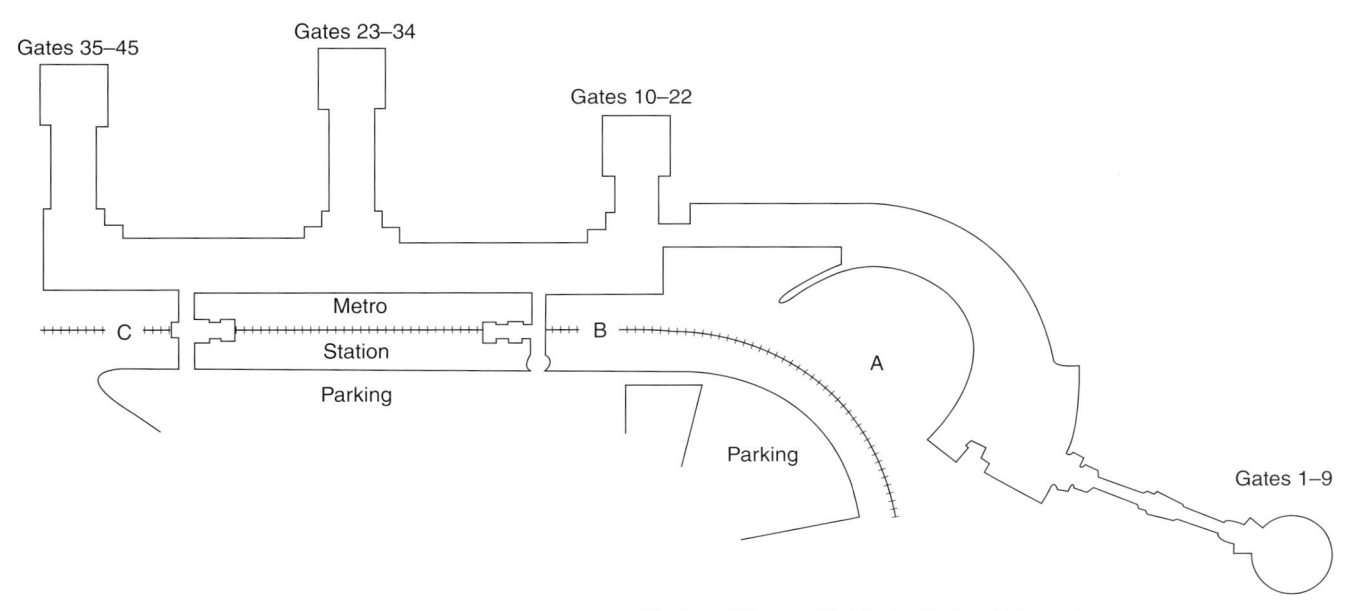

6.99 Ronald Reagan Washington National Airport diagram.

Transport terminals and modal interchanges

(a)

(b)

▲ 6.100 (a, b) Metro at airport terminal (Courtesy of Metropolitan Washington Airport Authority).

▲ 6.101 Metro leaving airport (Courtesy of Metropolitan Washington Airport Authority).

110　Airport/railway interchange: linked adjacent

6.3.8 San Francisco International Airport, USA

Interchange – metro railway terminus linked to five separate terminal buildings by people-mover.

The airport now handles 29.3 million passengers per year (ACI 2003 figure).

The airport has three domestic and one international terminals in a ring layout, all connected by AirTrain, a two-line system people-mover.

Rail access has been possible since June 2003 with the installation of AirTrain, which meets at one station (also a car park stop) with a terminus of the Bay Area Rapid Transit (BART) one level below. Furthermore, connection can be made with the national rail network at a station called Millbrae, one stop from the airport by BART.

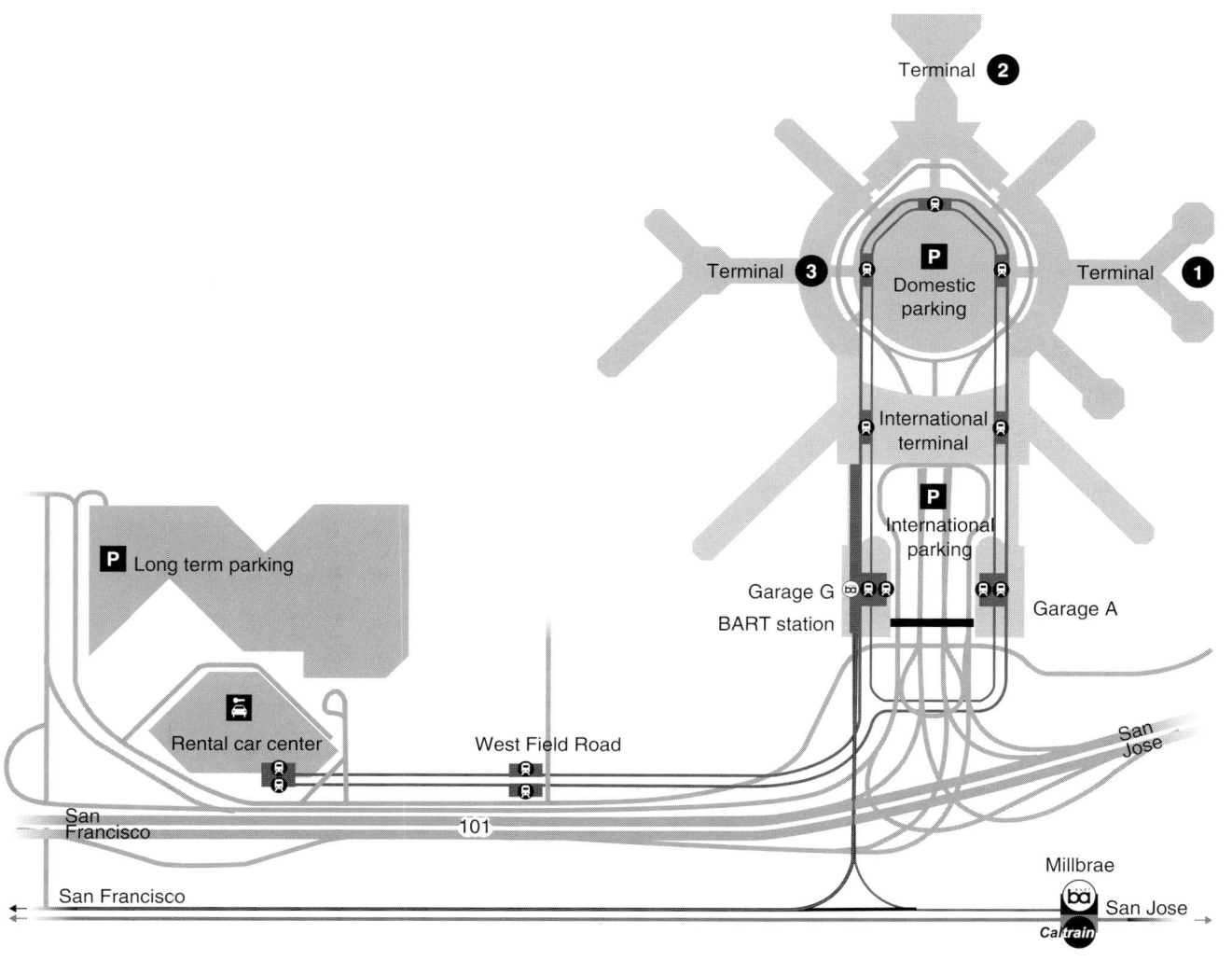

▲ 6.102 San Francisco International Airport: key map (Courtesy of San Francisco Airport).

Transport terminals and modal interchanges

6.103 Photo of San Francisco AirTrain (Courtesy of Bombardier).

Airport/railway interchange: linked adjacent

6.3.9 Inchon Airport, Seoul, South Korea

Interchange – multiple urban and national railway termini linked to one terminal, with more to come in future.

The airport has a capacity of 100 million passengers per year and the first phase, with capacity of 27 million passengers per year, was completed in 2001 in time for the 2002 World Cup. The reclaimed site is located between two existing islands 50 km from Seoul. For international traffic only, it replaces Kimpo Airport on the mainland much closer to the city of Seoul. The Transportation Centre is expected ultimately to serve 30 million passengers per year; 5000 car parking spaces are built in.

The fan-shaped centre incorporates five rail systems, for which platforms have been built in readiness: high-level light rail link to new International Business City, local rail (150 m platforms), national high-speed rail (400 m platforms), future inter-terminal rapid transit and, at the lowest level, a baggage system. The rail link to Kimpo with a 30-minute service is expected in 2005 and the main link to Seoul with a 50-minute service in 2007.

▲ 6.104 Inchon Airport, Seoul, South Korea: Transportation Centre, aerial view (Courtesy of Terry Farrell & Partners, architects).

> **TRANSPORTATION CENTRE TEAM**
>
> ARCHITECTS: Terry Farrell & Partners, with Samoo Architects and Engineers
> STRUCTURAL, MECHANICAL AND TRANSPORT ENGINEERS: DMJM
>
> **TERMINAL TEAM**
>
> ARCHITECTS: Fentress & Bradburn, with Korean Architects Collaborative

REFERENCES

Architecture Today, profile no. 1/Terry Farrell & Partners.

Dawson, S. (2002). Foreign exchange. *Architects Journal*, Metalworks supplement, Summer.

Glaser, K. (2002). Good Korea move. *Building Design*, 11 January.

▲ 6.105 Transportation Centre, night view (Courtesy of Terry Farrell & Partners, architects).

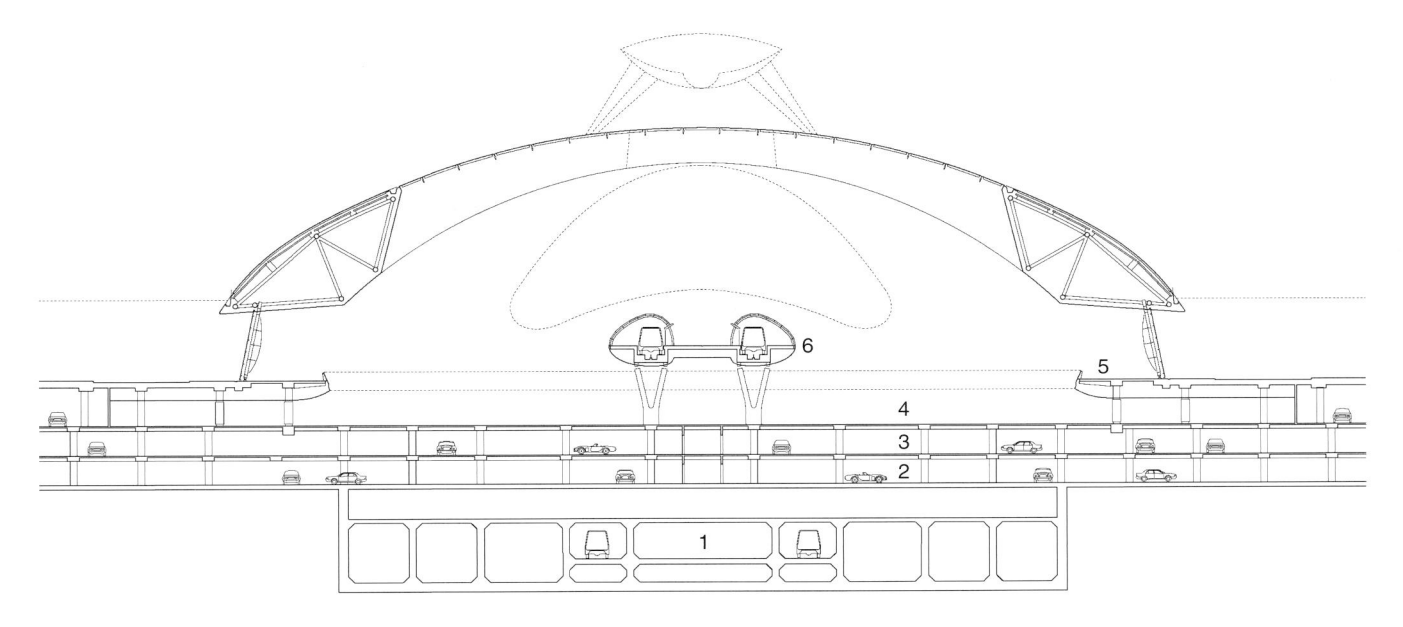

Key
1 Baggage handling
2 Platform and car park level 1
3 Car park level 2
4 Great Hall concourse, airport arrivals level and car park level 3
5 Ground level and links to buses, coaches and taxis
6 Light rail link to new city and airport departures

10 m

▲ 6.106 Transportation Centre, cross-section (Courtesy of Terry Farrell & Partners, architects).

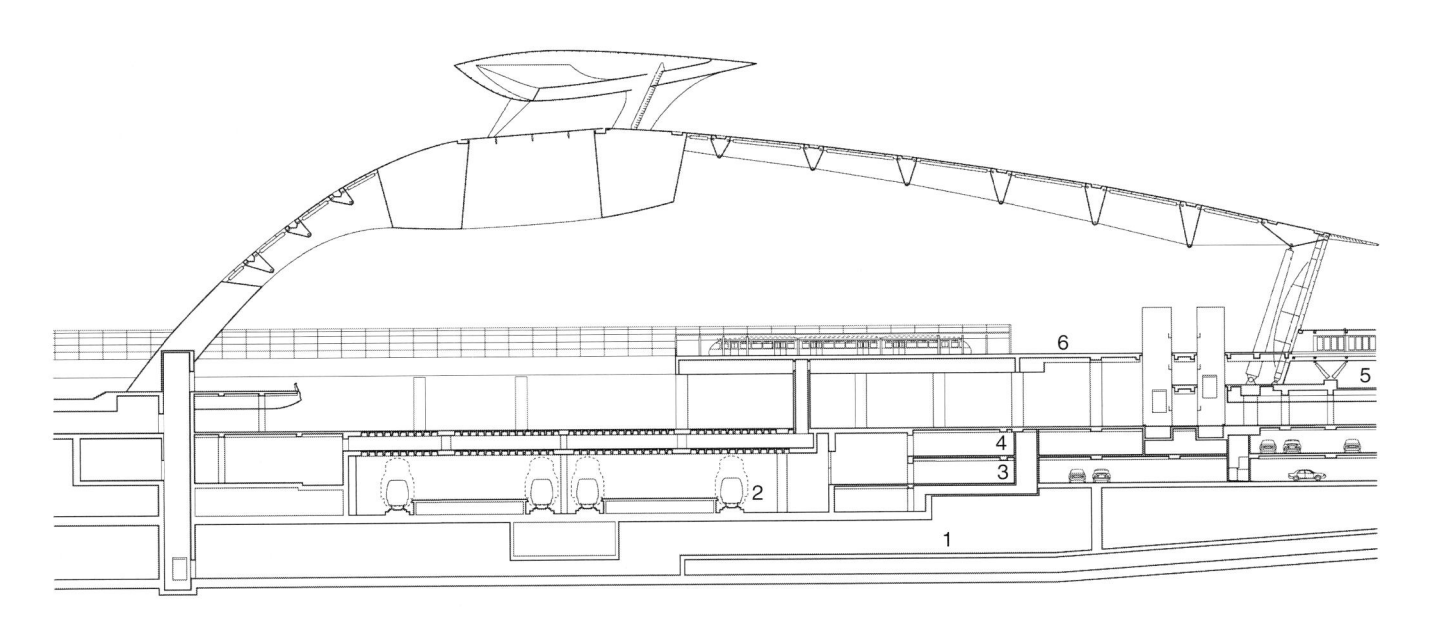

Key
1 Baggage handling
2 Platform and car park level 1
3 Car park level 2
4 Great Hall concourse, airport arrivals level and car park level 3
5 Ground level and links to buses, coaches and taxis
6 Light rail link to new city and airport departures

10 m

▲ 6.107 Transportation Centre, transverse section (Courtesy of Terry Farrell & Partners, architects).

6.4 Airport/railway interchange: remote

6.4.1 Luton Airport, UK

Interchange – national railway station linked to terminal building by 2 km shuttle bus route with potential for people-mover.

The airport now handles 5.5 million scheduled flight passengers per year (2002 figure) and this is attributable to the advent of 'no-frills' airline easyJet in 1995. Prior to this, the throughput of scheduled traffic was 10% of the 2002 figure.

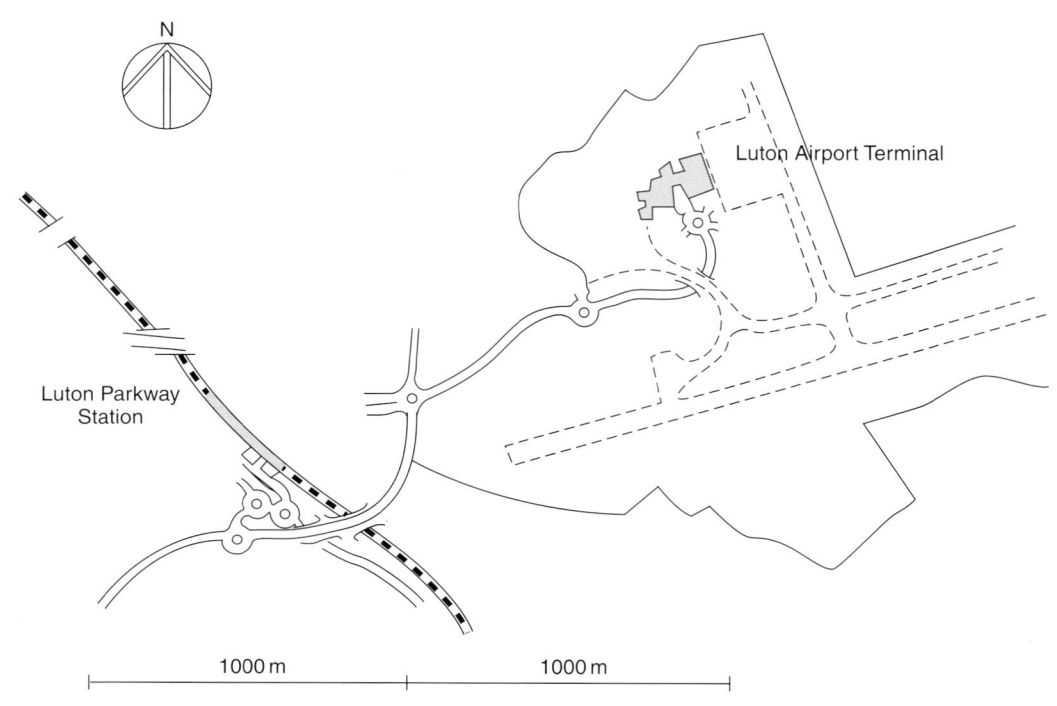

6.108 Luton Airport: map linking station and airport.

6.109 Shuttle bus at Luton Parkway Station.

Transport terminals and modal interchanges

6.110 Shuttle bus at airport terminal building, showing new Departures Terminal, 1999, built to concept design by Foster & Partners.

6.111 Heathrow bus at airport terminal building.

Facilities were improved in 1999 with the opening of the new Departures Terminal illustrated in Figure 6.110, together with the new eastern apron.

Rail access has been possible since 2001 with the construction of Luton Parkway Station on the main line from St Pancras to the Midlands and served by Thameslink trains running through London from north to south; a 2-kilometre bus ride takes passengers from rail station to airport terminal.

The airport's target of 34% passenger use of public transport was not achieved in 2002: 7% of passengers used bus/coach (excluding the railway station shuttle bus) and 17% used the rail link.

REFERENCES

Kehoe, P. (2003). Presentation at Future Terminal Conference, London.
London-luton.co.uk/pdf/download website (2003). Surface access strategy, July.

Airport/railway interchange: remote **117**

6.4.2 Düsseldorf Airport, Germany

Interchange – national railway station linked to car park and terminal building by 2.5 km people-mover.

The airport now handles 14.7 million passengers per year (2002 figure) and, after a major construction programme to designs by J.S.K. Perkins & Will following a fire in 1996, has a present capacity of 22 million passengers per year.

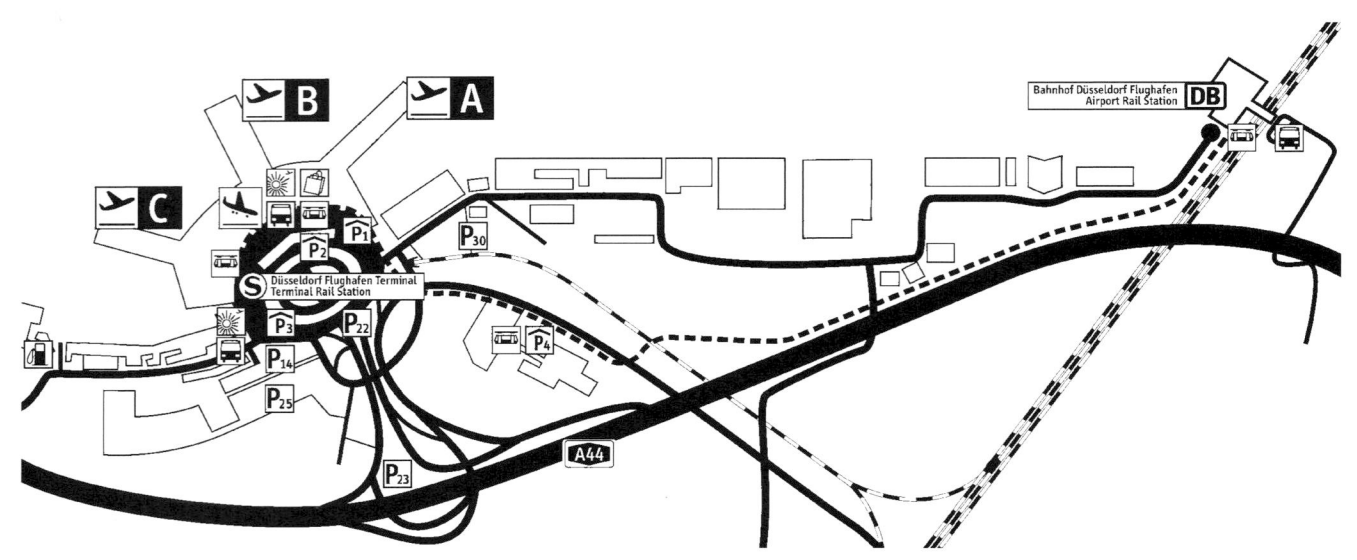

6.112 Düsseldorf Airport: key map (Courtesy of Düsseldorf International).

6.113 Airport aerial photo (Courtesy of Düsseldorf International).

Transport terminals and modal interchanges

6.114 Aerial photo of rail station (Courtesy of Düsseldorf International).

6.115 Photo of rail station in action (Courtesy of Düsseldorf International).

Transport terminals and modal interchanges

This airport is 50% owned by the City of Düsseldorf, 30% by Hochtief and 20% by Aer Rianta.

Rail access has been possible since 2003 with the installation of Skytrain, a 2.5-kilometre suspended monorail people-mover, which runs almost literally through the terminal. The capacity of the system is 2000 passengers per hour in each direction. The main line rail station offers 300 connections per day. This station incorporates a check-in terminal with 20 desks.

TEAM

CLIENT: Flughafen Düsseldorf International
ARCHITECTS: JSK International Architekten und Ingenieure GmbH

▶ 6.116 Photo of Skytrain leaving rail station (Courtesy of Düsseldorf International).

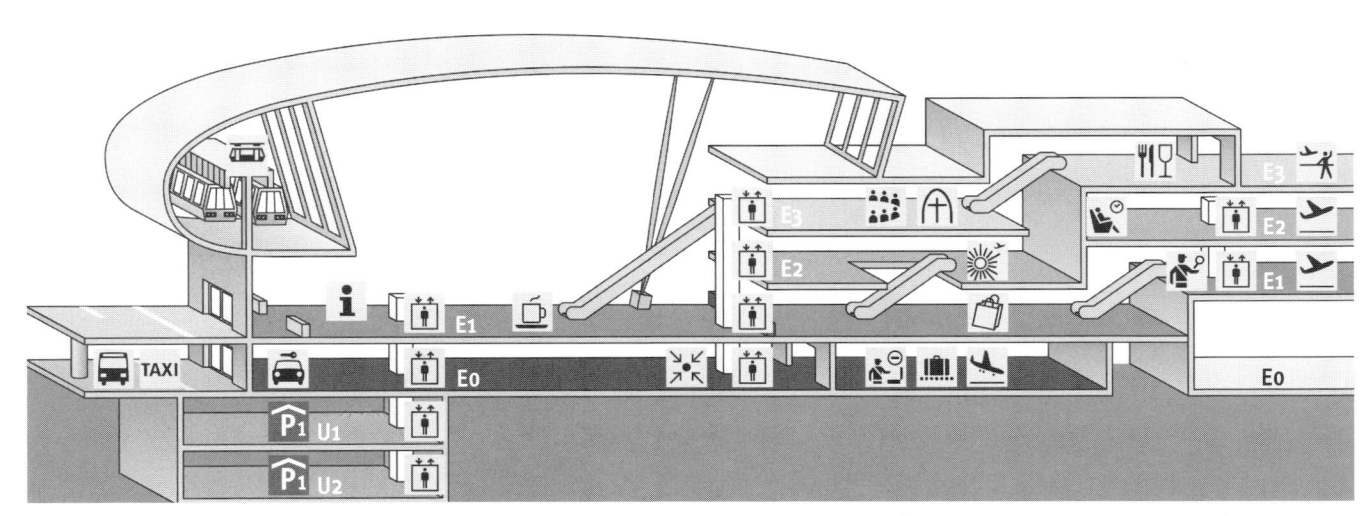

▲ 6.117 Terminal cross-section (Courtesy of Düsseldorf International).

120 Airport/railway interchange: remote

6.118 Internal photo of terminal concourse (Courtesy of Düsseldorf International).

Transport terminals and modal interchanges

6.119 Photo of Skytrain 'inside' terminal (Courtesy of Düsseldorf International).

122 Airport/railway interchange: remote

6.5 Multiple railway station/bus and coach/car interchanges: vertical separation

6.5.1 Lyon Perrache Railway Station, France

Interchange – metropolitan railway station with trams below and buses and coaches and national rail above, a megastructure.

Lyon is France's second largest city, with a population of over 1 million. For many years the city has boasted a four-line metro system and excellent bus services. Nevertheless, in the face of serious traffic congestion, it is the first French city to adopt an urban transport plan. In 2000 the first two of a series of tramway lines were opened, based around the Perrache railway and metro station. The trams travel and stop beneath the rail and bus station and alongside the metro terminus in a redundant underpass.

The location of the exceptional metropolitan interchange coincides with the junction of two national motorways, the A6 and A7, the latter passing through the structure of the interchange, with buses circulating above and trams below.

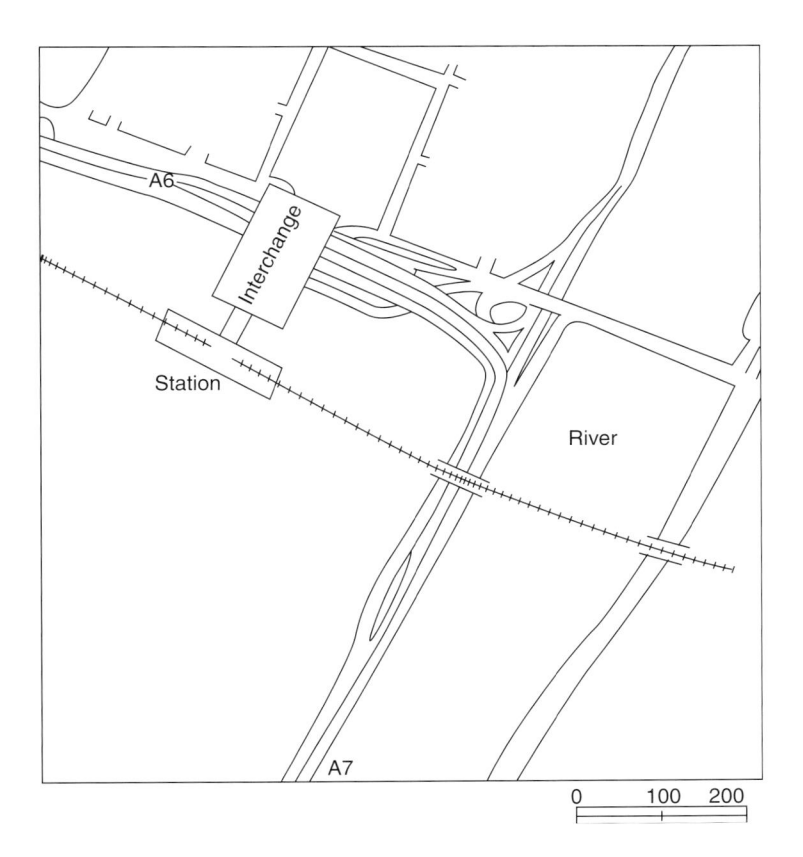

▶ 6.120 Lyon Perrache Railway Station: key map.

Transport terminals and modal interchanges

▶ 6.121 Lyon Perrache tram line towards Pont Gallieni.

▶ 6.122 Lyon Perrache tram turning adjacent to rail station.

6.5.2 Circular Quay Interchange, Sydney, Australia

Interchange – metropolitan railway station with buses and ferry jetties below.

This is one of the most dramatic scenic metropolitan interchanges in the world, with the famous Sydney Opera House on one side and the towering Harbour Bridge on the other. At the head of the inlet known as Sydney Cove, the historic focal point of this harbourside city, is the Circular Quay ferry terminal. The ferry terminal has five jetties from which picturesque ferries serve the 37 other wharves and jetties of Sydney Harbour as well as sightseeing cruises. Above the concourse is the metropolitan railway station with direct services to most parts of the city and the southern suburbs. One change of train gives access by train to all the northern suburbs by way of the famous Harbour Bridge. Buses stop at the street level concourse and serve all parts of the inner city. Cruise liners also dock acjacent at the Overseas Passenger Terminal (see Chapter 6.7.3).

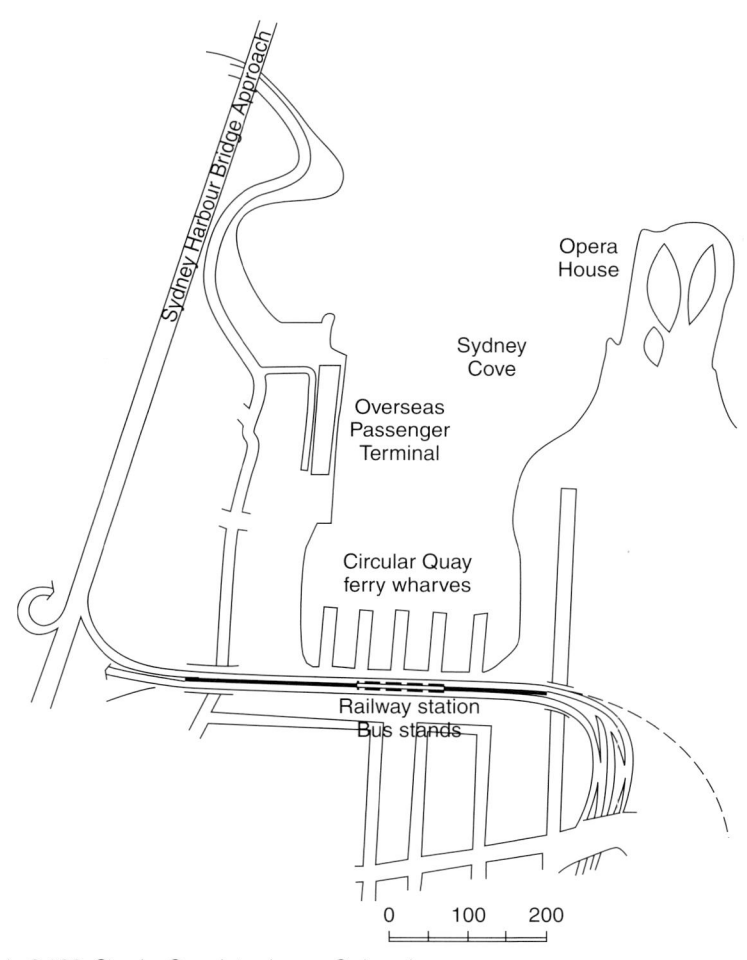

6.123 Circular Quay Interchange, Sydney: key map.

Transport terminals and modal interchanges

6.124 View of ferry terminal with railway station above.

6.6 Multiple railway station/bus and coach/car interchanges: contiguous

6.6.1 Ashford International Station, Kent, UK

Interchange – international railway station with domestic railway station, buses and coaches and car park adjacent.

This location was chosen well before the opening of the Channel Tunnel and the availability of through Eurostar trains from London and Paris or Brussels. The advantages of locating what is partly a park-and-ride interchange at Ashford Station were twofold: the regeneration of the town, and the offering of connections with local and regional trains. The attractions to residents of Kent and Sussex of a 2000-space car park with immediate access to trains to Paris are evident.

In the terminal or station building, domestic as well as arriving and departing international passengers must be segregated once they have passed the ticket barrier, so separate routes are provided to separate platforms. Domestic passengers have a dedicated subway, facilities for departing international passengers are at the upper level and they have a bridge to the dedicated 'international' island platform, and arrivals use a subway which brings them into the immigration area of the terminal. On the far side of the tracks from the terminal is a high-speed bypass route for non-stop trains.

REFERENCE

Architecture Today (1996). Ashford. Issue 68, May.

Transport terminals and modal interchanges

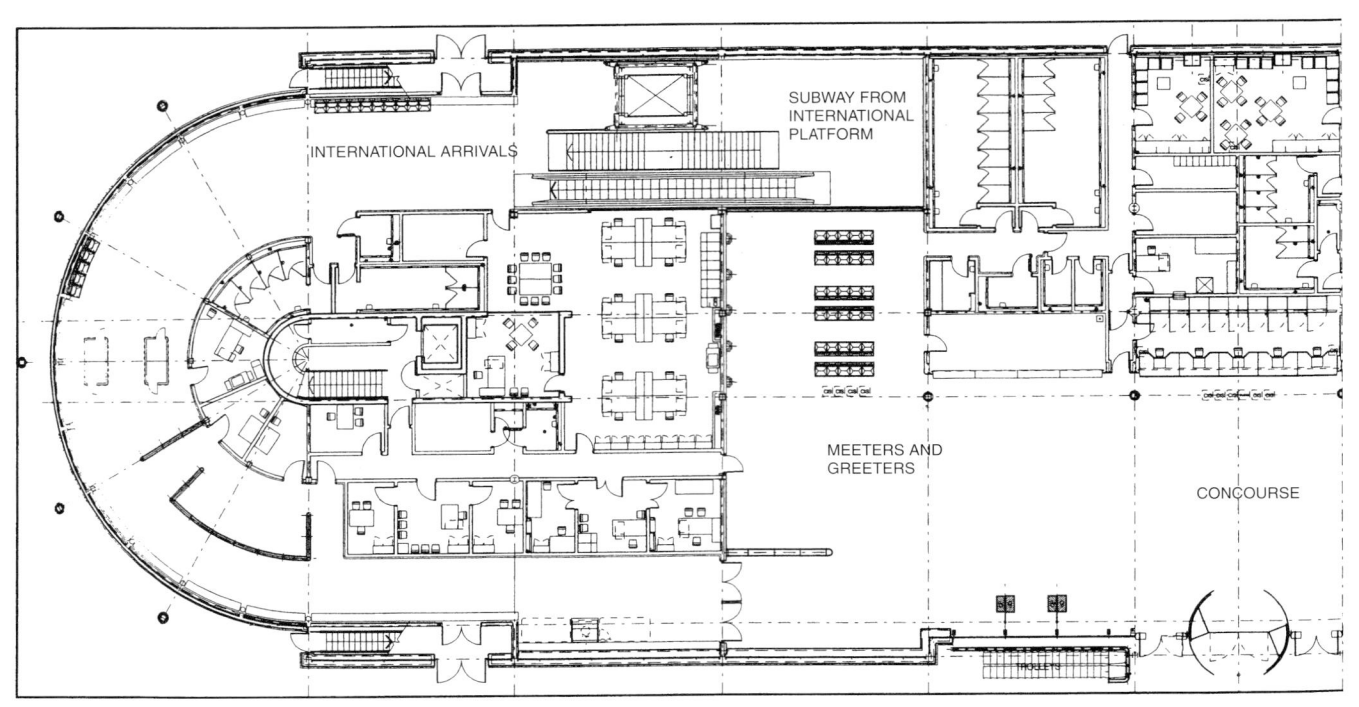

▲ 6.125 Ashford International Station, Kent: terminal building plan (Courtesy of Jacobs, in association with Nick Derbyshire Design Associates).

▲ 6.126 Railway-side photo showing bypass for non-stop trains, domestic platform, international station and terminal building in background.

Transport terminals and modal interchanges

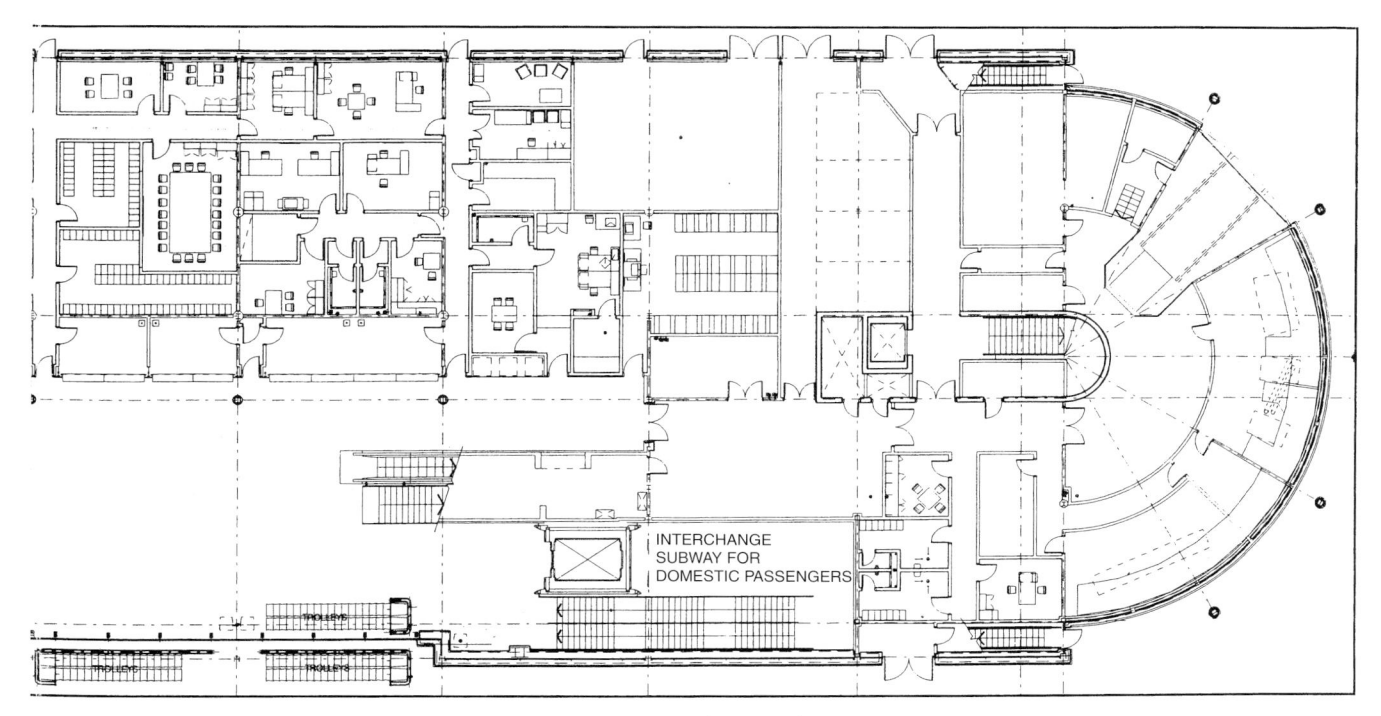

▲ 6.125 (Continued)

▶ 6.127 Main entrance.

Multiple railway station/bus and coach/car interchanges: contiguous 129

Transport terminals and modal interchanges

6.128 Ashford coach park at south end of terminal building.

6.129 Panorama showing car park and bridge to terminal building, Ashford.

6.6.2 Channel Tunnel Terminal, Cheriton, Kent, UK

Interchange – terminus for dedicated rail service carrying road vehicles exclusively.

As soon as the basic decision was made in the early 1980s to build a rail-only twin-tunnel under the English Channel, the need for two unusual giant transit stations was established. When the tunnel opened in May 1994, the amenities, the border controls and, most important of all, the loading ramps were complete, to complement the rail trackwork to enable dedicated trains to turn round. The project illustrated here is at the English end, just outside Folkestone in Kent.

The commercial service for the carriage of cars started in December 1994. After carrying 1.2 million cars, motorcycles and caravans and 23 000 coaches in the first year, annual traffic has averaged 2.5 million cars, motorcycles and caravans and 75 000 coaches over the eight-year period from 1996 to 2003. Thus, it is estimated that the amenities are used by 3.5 million departing passengers per year.

The key to passenger handling is that car passengers remain in their vehicles for the 35-minute journey. Each train carries 180 cars or a combination of cars and coaches. A different train and system provides for heavy goods vehicles.

REFERENCE

Byrd, T. (1994). *The Making of the Channel Tunnel*. New Civil Engineer/Transmanche Link publication.

▶ 6.130 Aerial view from the south, with arriving train on the left, departing train on the right (Courtesy of Eurotunnel plc).

Transport terminals and modal interchanges

▶ 6.131 Channel Tunnel Terminal, Cheriton, Kent: aerial view from the north (Courtesy of Eurotunnel plc).

KEY

T   toll booths
S   services
F   frontier controls
AP allocation zone (passenger vehicles)
AF allocation zone (freight vehicles)
L   loading area (with bridges)
U   unloading area (with bridges)
R   returning train loop

6.6.3 Manchester Piccadilly Station, UK

Interchange – metropolitan railway terminus with buses adjacent and urban light rail below.

Winner of the UK Integrated Transport Award 2003 Large Interchange Project of the Year.

Piccadilly is one of three main stations in Manchester and handles over 55 000 passengers and 1000 train movements every day.

Following upgrading in 2002 in time for the Commonwealth Games, it is now regarded as one of the best stations and interchanges in the UK.

Escalators lead directly from the platform-level concourse to a new entrance in Fairfield Street. This serves taxis, private cars picking up and setting down, southbound and eastbound buses and the southbound Metrolink light rail, and free shuttle buses to the city centre have exclusive use of a

6.132 Manchester Piccadilly Station concourse (Courtesy of Building Design Partnership, Manchester).

Transport terminals and modal interchanges

▲ 6.133 Concourse with trains (Courtesy of Building Design Partnership, Manchester).

ramp to concourse level which was previously congested and used by cars and taxis. Unfortunately, it was not possible to relocate the nearby coach station at the Fairfield Street entrance/interchange.

TEAM	
CLIENT:	Network Rail
ARCHITECTS:	Building Design Partnership and EGS Design
ENGINEERS:	URS Thorburn Colquhoun
TRAFFIC MANAGEMENT:	Faber Maunsell

REFERENCE

Architecture Today (2003). Train of thought. Issue 141, September.

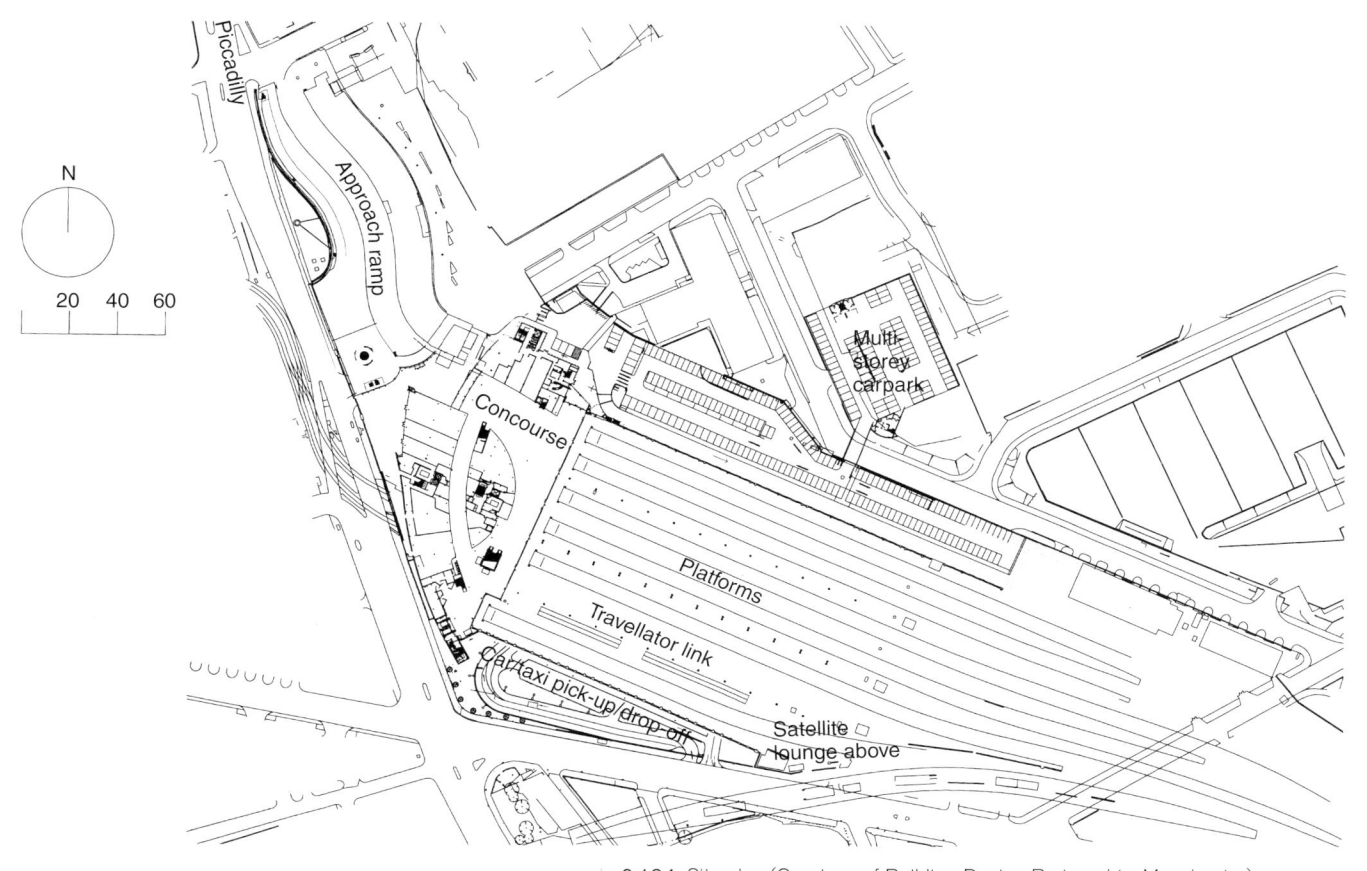

6.134 Site plan (Courtesy of Building Design Partnership, Manchester).

Transport terminals and modal interchanges

6.6.4 Stratford Station, London

Interchange – urban railway station and metro and light rail terminus with buses adjacent. Future international rail station adjacent (see section 6.6.5).

This station has long been an important London-edge interchange point between two rail lines, Central Line tube and buses. With the advent of the Docklands Light Railway (DLR) northern terminus and the eastern terminus of the Jubilee Line tube in the last 15 years, the complexity has increased and passenger convenience has been compromised.

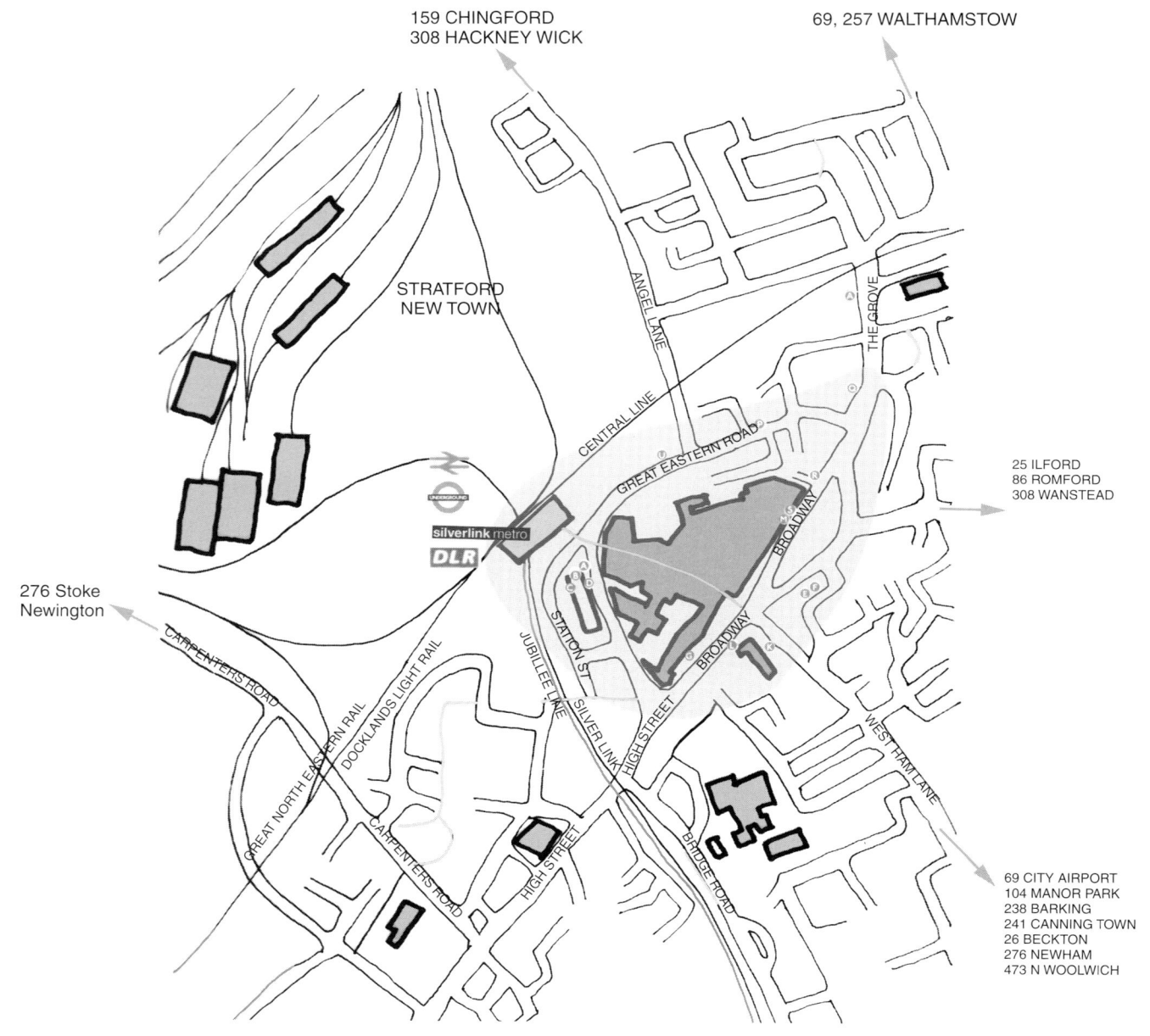

6.135 Stratford Station, London: diagrammatic area map (Source: study by RCA/Scott Brownrigg for Helen Hamlyn Research Centre).

136 Multiple railway station/bus and coach/car interchanges: contiguous

Transport terminals and modal interchanges

▲ 6.136 Panoramic view (Source: study by RCA/Scott Brownrigg for Helen Hamlyn Research Centre).

▲ 6.137 Main view of new station (Source: study by RCA/Scott Brownrigg for Helen Hamlyn Research Centre).

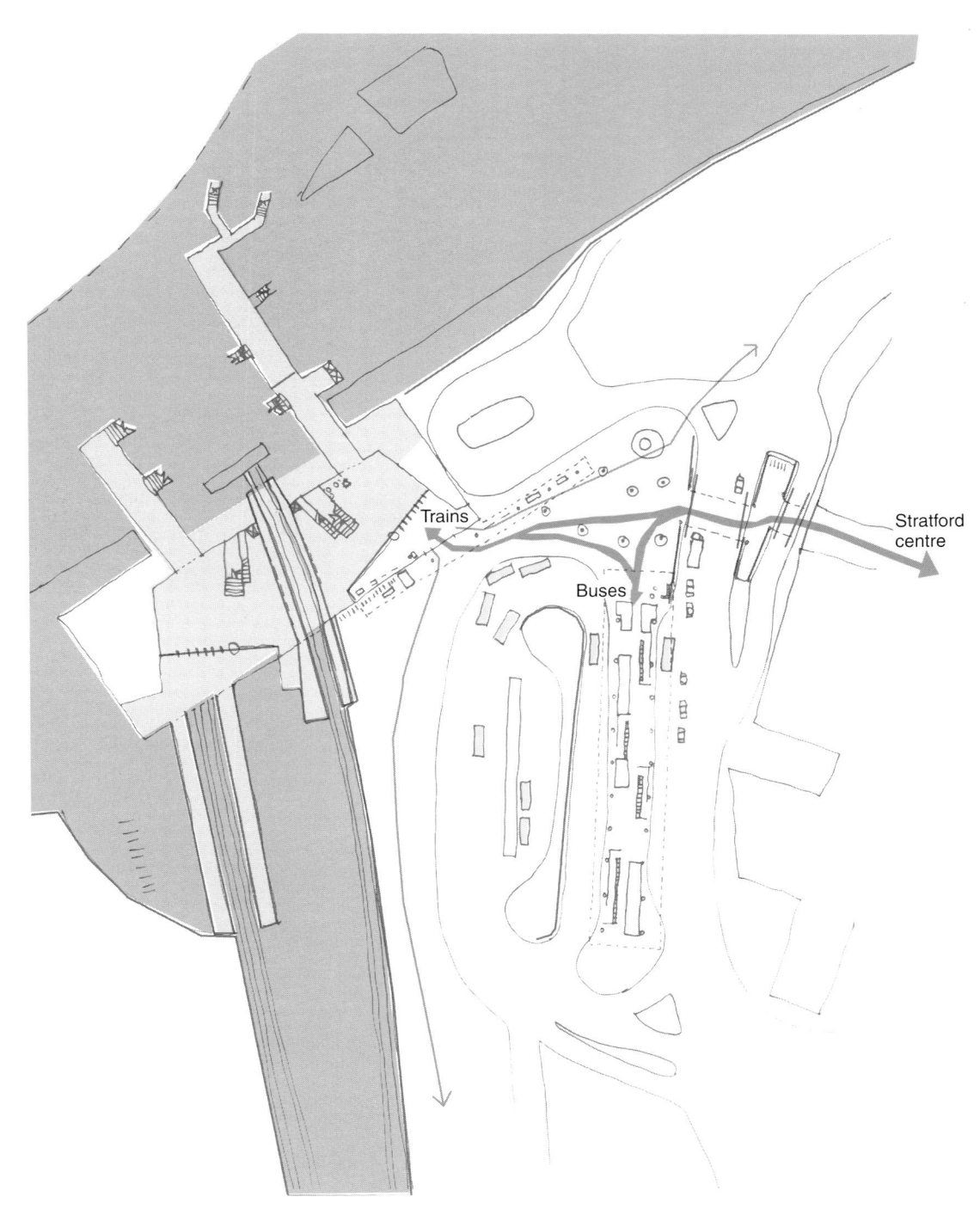

▲ 6.138 Ground-level plan (Source: study by RCA/Scott Brownrigg for Helen Hamlyn Research Centre).

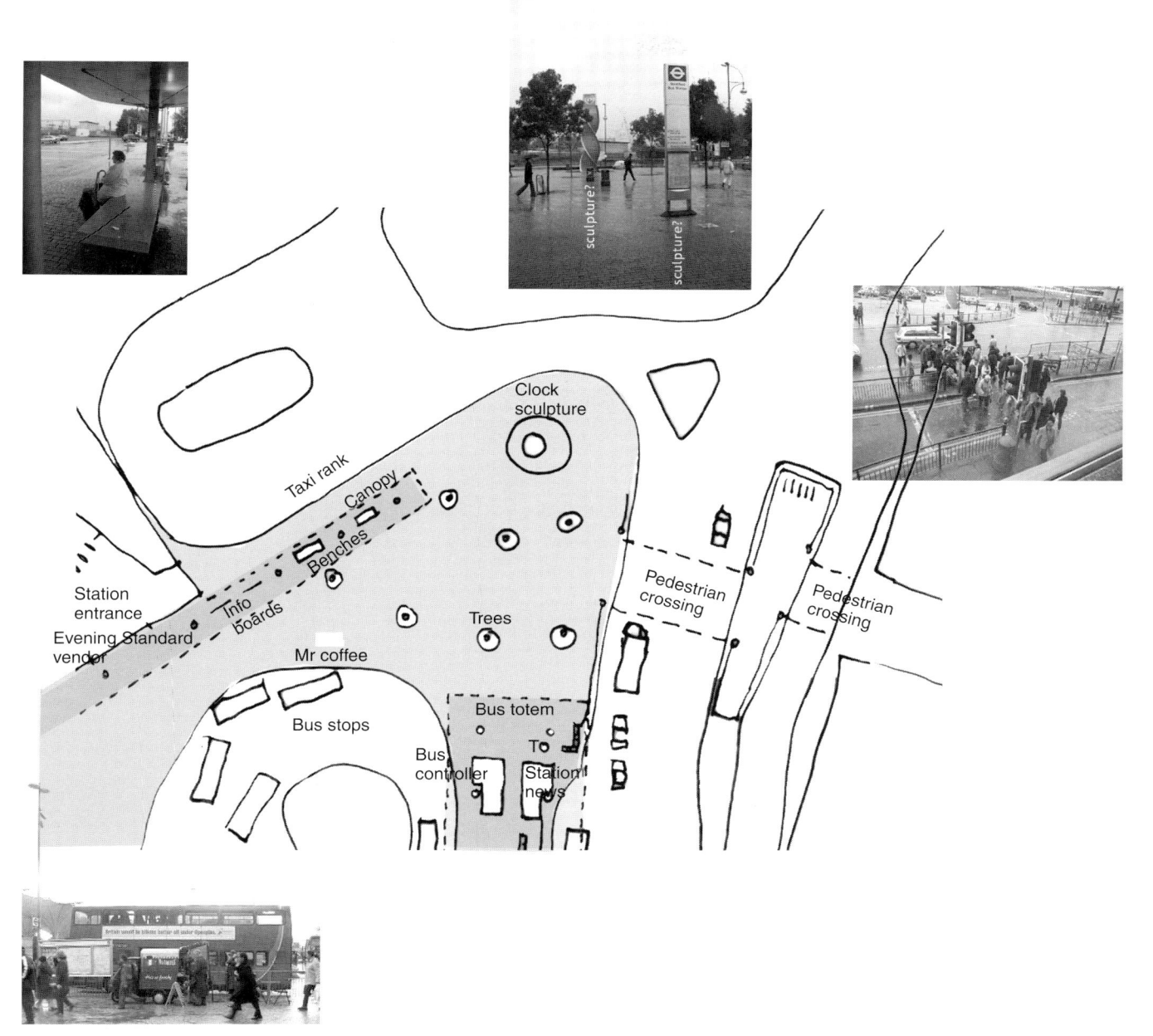

Transport terminals and modal interchanges

▲ 6.139 Station forecourt plan (Source: study by RCA/Scott Brownrigg for Helen Hamlyn Research Centre).

The new station building, essentially built for the Jubilee Line terminus, nevertheless links the subways which feed the rail and DLR platforms at high level with the other passenger routes. Bus passengers are the poor relations.

In fact, one of the studies described in Chapter 2 as exploring new standards and expectations relates to the generic features of an urban interchange like Stratford.

Multiple railway station/bus and coach/car interchanges: contiguous

Transport terminals and modal interchanges

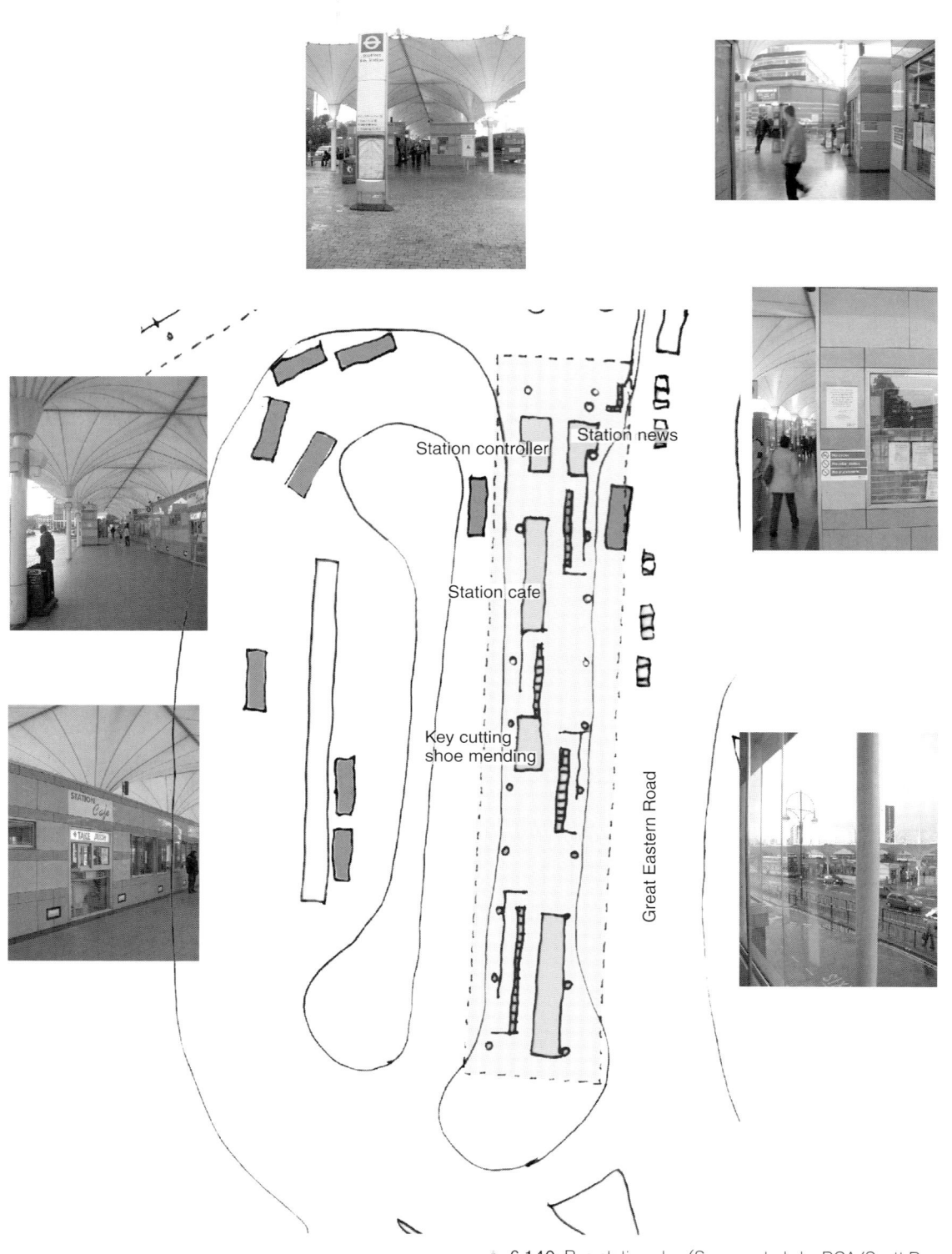

Station controller
Station news
Station cafe
Key cutting shoe mending
Great Eastern Road

▲ 6.140 Bus station plan (Source: study by RCA/Scott Brownrigg for Helen Hamlyn Research Centre).

140 Multiple railway station/bus and coach/car interchanges: contiguous

TEAM, 1994 REDEVELOPMENT

CLIENT: London Underground
ARCHITECTS: Wilkinson Eyre

REFERENCE

Scott, F. (2003). *InterchangeAble*. Helen Hamlyn Research Centre, Royal College of Art, London.

6.6.5 St Pancras Station, London

Interchange – international railway terminus and multi-line metro station with buses adjacent.

The Eurostar service is due to operate in 2007 from London St Pancras to Paris, Amsterdam and Brussels via Stratford (Chapter 6.6.4), Ebbsfleet and Ashford (Chapter 6.6.1). The operator believes that traffic will be attracted away from airlines, even to the extent of removing the need for 250 000 short-haul flights per year.

The selection of the original and notable St Pancras Station as the terminus for the Channel Tunnel Rail Link (CTRL) ensured preservation and extension of a famous Victorian building. The 'train-shed' terminus of the Midland Railway, which opened in 1868 with a design by W.H. Barlow, the Engineer-in-Chief of the Midland Railway, was in its time the largest undivided space ever enclosed. The fact that it is fronted by what was the ornate Midland Grand Hotel, opened in 1873 with a design by Sir George Gilbert Scott but closed in 1935, destined the hotel to be identified as the station and destined the train-shed to be largely unrecognised. However, the extension northwards to cater for the 400-metre-long international trains will ensure its long-term recognition within a landmark transport interchange. The new train-shed will accommodate CTRL trains as well as Midland Mainline trains serving cities like Leicester and Nottingham. Allowance is also made for a new terminus for the Heathrow Express, providing a high-speed, high-quality link to Heathrow Central (Chapter 6.3.4), Heathrow Terminal 4 (Chapter 6.1.5) and Terminal 5 (Chapter 6.1.4). The former hotel, now St Pancras Chambers, remains as a landmark.

Connections to the four sets of platforms of the London Underground system as well as numerous bus and coach routes are incorporated in the redevelopment project, together with new sub-ground platforms for the relatively new north–south Thameslink rail line, which weaves its way through London.

To quote from Andrew Lansley, lead architect with RLE:

> St Pancras gives us architecture free of charge, with W.H. Barlow's Grade 1 listed shed. Wherever we touched the building we had to make it look as if Barlow had not let go of it ... But we are also being quite radical. We are cutting into the floor of the shed to bring light to the 'undercroft' ... For international trains to compete with short-haul flights, we realised St Pancras had to be planned for circulation efficiency as well as for good passenger facilities. Thus it is possible to board a train within five minutes of arrival. The international capacity (one third of the total) equates to the numbers using Heathrow Terminal 4.

TEAM

CLIENT:	Union Railways (part of London and Continental Railways) for Channel Tunnel Rail Link (CTRL)
CONCEPT ARCHITECTS:	Foster & Partners
ARCHITECTS:	Rail Link Engineering (RLE) – a consortium of Arup, Bechtel, Halcrow and Systra

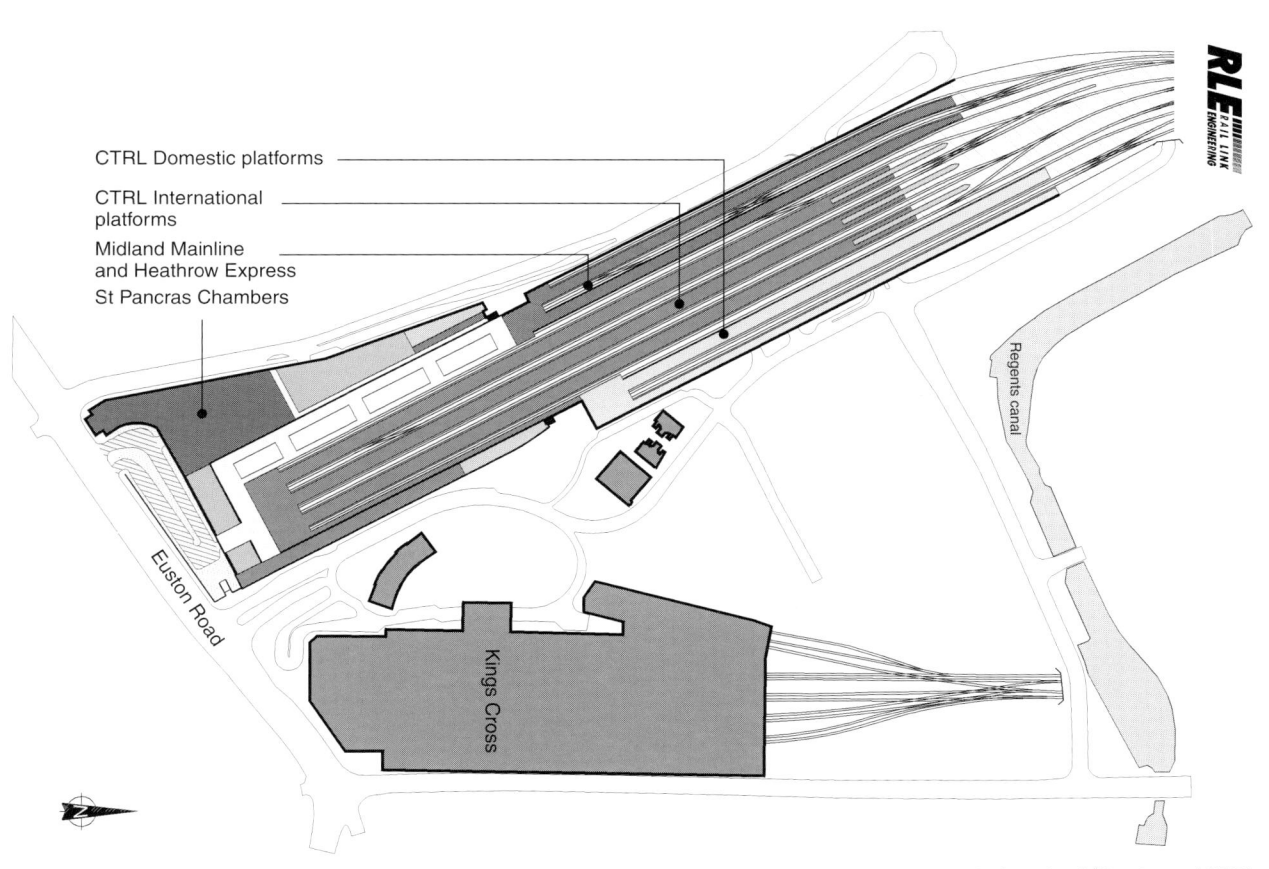

▲ 6.141 St Pancras Station, London: diagram, platform level (Courtesy of CTRL Press Office).

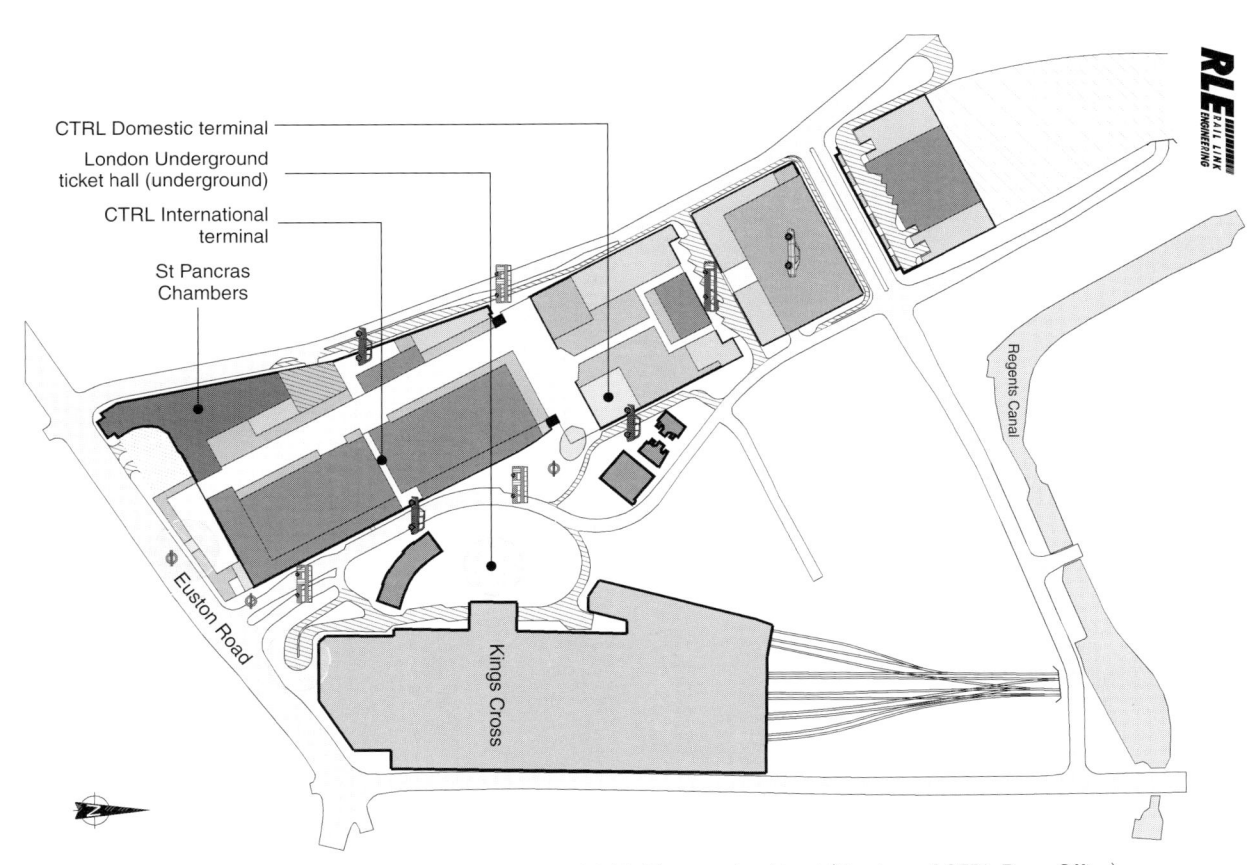

▲ 6.142 Diagram, street level (Courtesy of CTRL Press Office).

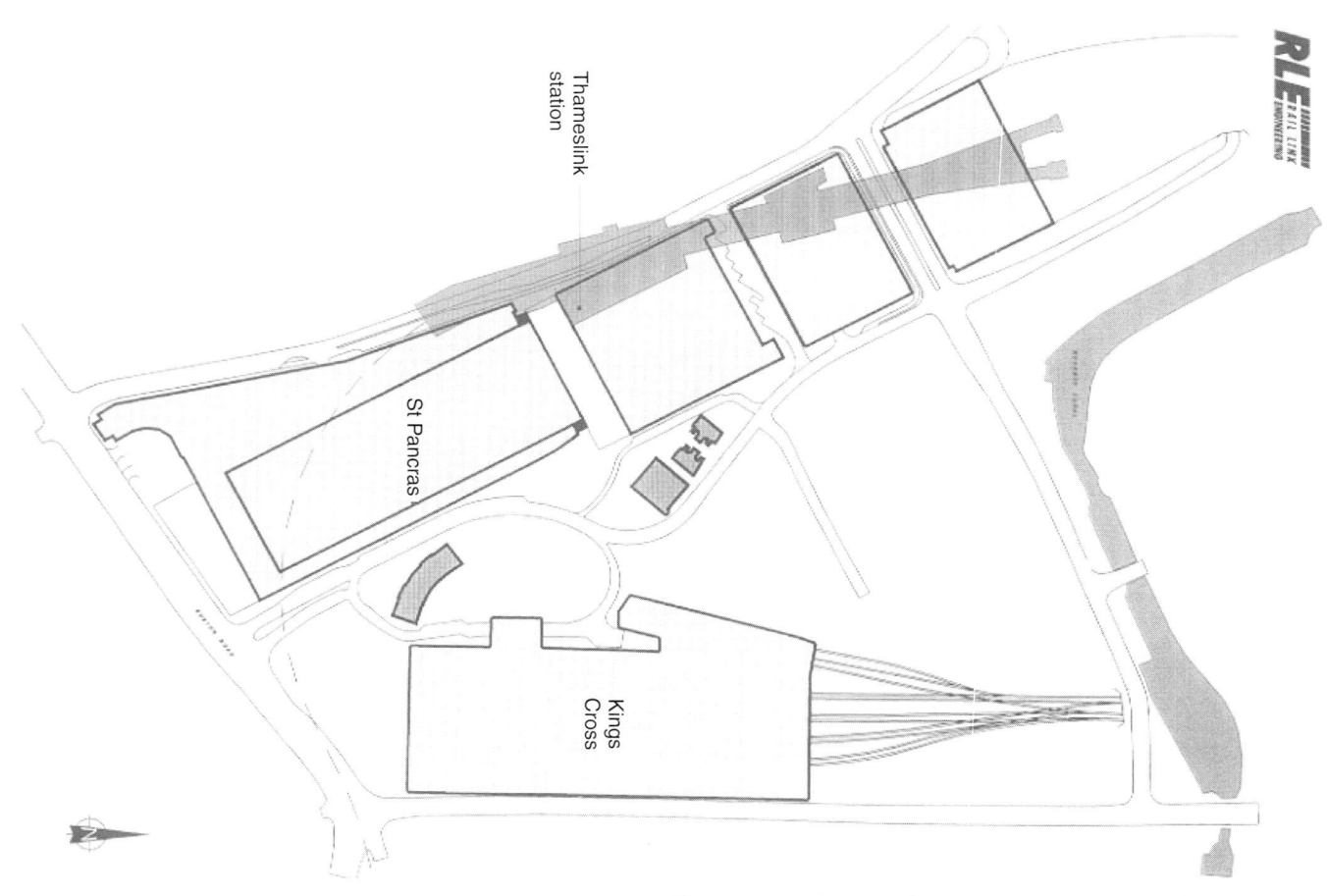

▲ 6.143 Diagram, subsurface level (Courtesy of CTRL Press Office).

▲ 6.144 Interior view inside train-shed (Courtesy of CTRL Press Office).

Transport terminals and modal interchanges

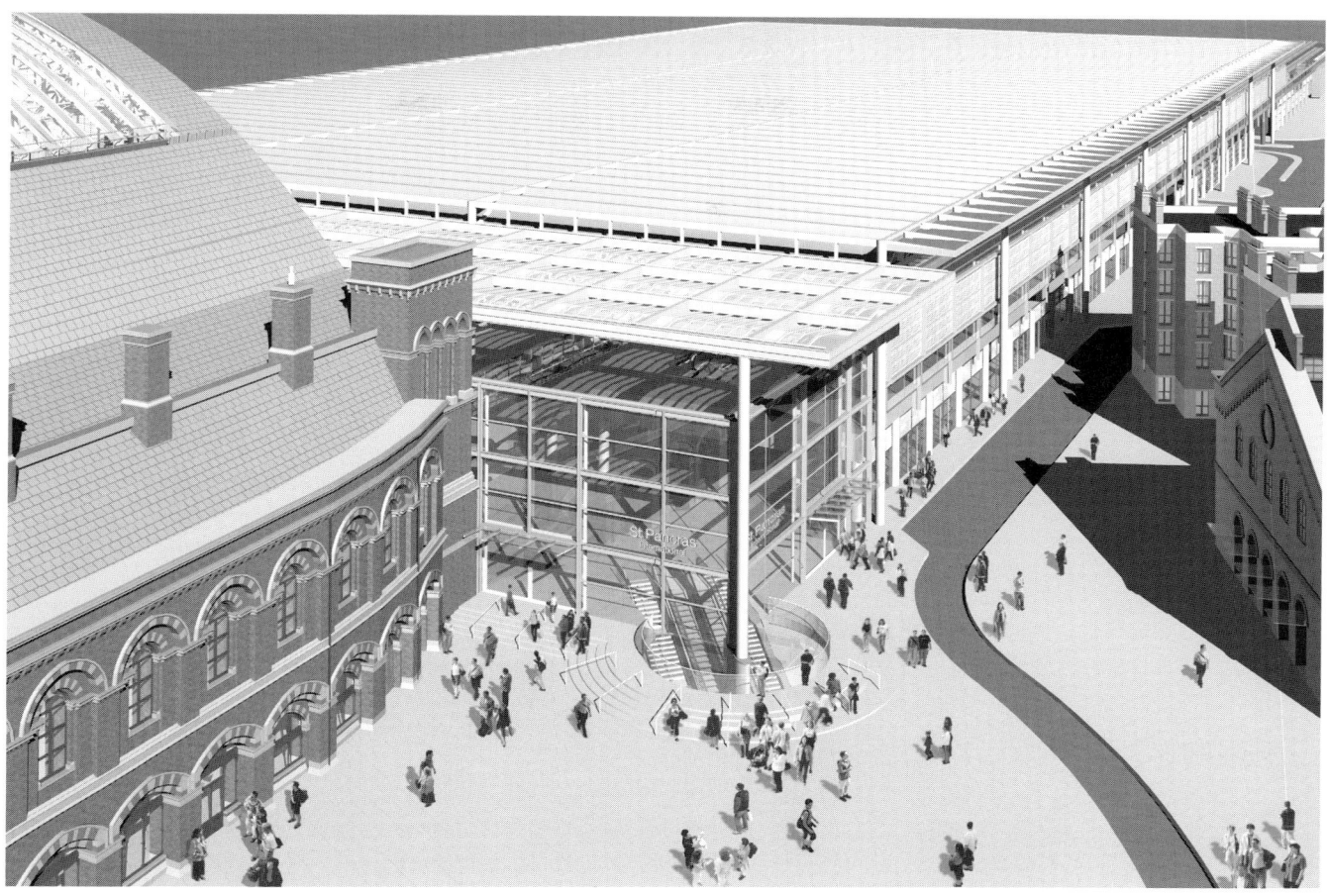

6.145 Exterior main entrance view (Courtesy of CTRL Press Office).

6.146 Coach road view (Courtesy of CTRL Press Office).

REFERENCES

Marston, P. (2003). Eurostar aims to ease congestion. *Daily Telegraph*, 23 June.
Young, E. (2004). You wait for ages, then .... *RIBA Journal*, February.

Multiple railway station/bus and coach/car interchanges: contiguous **145**

6.6.6 Enschede Station, The Netherlands

Interchange – urban railway station with buses adjacent.

Award winner: International Integrated Interchange of the Year, 2003.

This description is based on a paper prepared by Michael Stacey of Brookes Stacey Randall, architects.

The project has two stages: the first is the reconstruction of the bus station and the second the refurbishment of the train station, which was built in 1952. The overall aims of the project are to:

- Represent the city of Enschede and promote regeneration.
- Create a better and easily used transport interchange.
- Accommodate a high-quality public transport route, guided bus or tramway.

The project was partly funded by the European Social Fund.

This urban intervention has been designed to create a new civic identity for the Municipality of Enschede and a transport interchange of simple functionality. The bus station is based on the use of information technology to minimise the physical construction required and facilitate interchange. We studied the most logical routes from train to bus and bus to train, seeking a balance between moving, standing, orienting, looking, being seen, saying goodbye, seeking information and greeting a friend. Part of the challenge of the project was to relate the large-scale need for buses turning with the scale of a walking person, creating spaces for people to wait and meet and new public spaces forming an entrance to the city of Enschede.

Landscaping

The hard landscaping was of critical importance and is formed of three primary elements: precast concrete tiles in stainless steel frames, tarmac and granite elements. These are used to delineate the areas for pedestrians and for vehicle movement. The square in front of the Stoa is differentiated as a pedestrian square by the generous planting of trees, each of which is uplit from flush ground-based fittings. As the trees mature, this space should become even more inviting and provide 'natural' shelter for people to wait in.

Display masts

These are conceived as information 'mirrors' focused on the waiting traveller under the Stoa. The displays are held in position by a mast which is designed to be like a human backbone, intended to be read as sculpture and containing the nervous system – the cables and wires that feed the displays. The displays are electronically controlled 'flip dot' destination boards. The displays are scaled to communicate across the space necessary for the buses and are used to create a humanising scale.

Stoa

The Stoa is designed as a waiting space set in a safe and pleasing environment; full shelter is provided in the train station itself. The design and proportions of the steel frame of the Stoa are inspired by the precast concrete frame of Schelling's train station built in 1952, which was refurbished in stage 2. The Stoa has a clearly articulated glass roof, which is supported on aluminium bronze castings. This roof appears to float above uninterrupted space. The castings are intended to be sculptural elements which can be enjoyed in themselves and contribute to a progressive lightening of the

Transport terminals and modal interchanges

structure to end in the fine glass edge. In essence, information technology has been used to minimise the need to build physical enclosures; a waiting passenger will be able to choose to wait in the station or under the Stoa, watch the choreographed arrival of their bus on a display screen and then cross to the boarding point.

Enschede Interchange is a practical example of an integrated transport interchange, which accommodates trains, trams, bicycles, cars and buses. This is set in a series of pleasing public spaces; together it forms a vital access and presentational tool for the city.

Rail station

Structural analysis of the 1950s concrete frame of the station established that it was possible to retain the majority of this structure following selective repairs. The café was relocated to a new entrance directly addressing the public squares which form the bus station. This is covered by a generous glass canopy and takes the form of a bridge link over a dedicated cycle route. This is both a through route and leads down to the undercroft of the station, which provides 'vast' enclosed cycle storage, which is monitored by a cycle repair shop at its entrance. If a cyclist does not want to pay the modest charges of this secure and dry location, plentiful cycle racks have been provided at the level access points to the station.

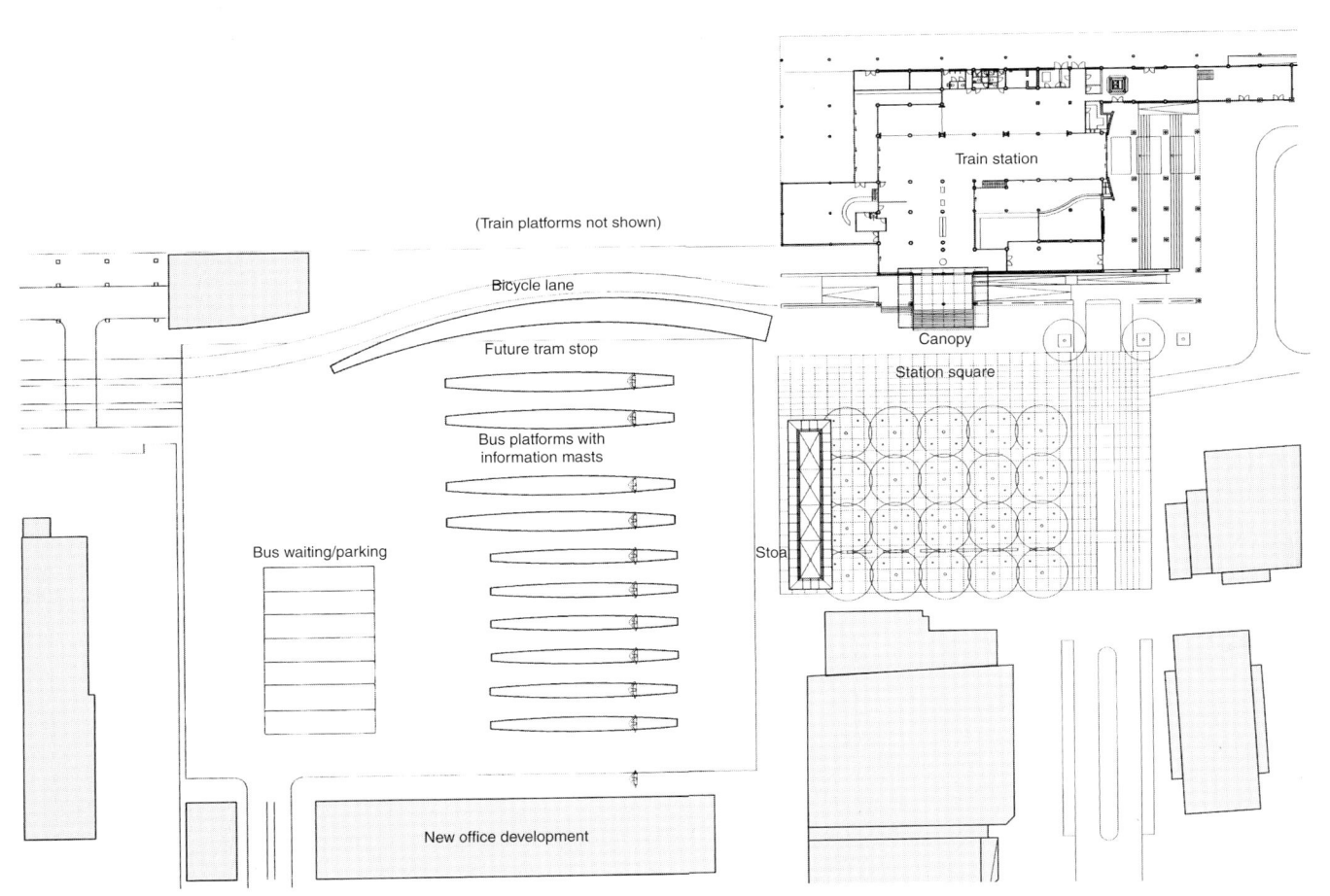

▲ 6.147 Enschede Station, The Netherlands: site plan (Courtesy of Brookes Stacey Randall).

Multiple railway station/bus and coach/car interchanges: contiguous

Transport terminals and modal interchanges

▶ 6.148 View of bus platform with information masts (Courtesy of Brookes Stacey Randall).

TEAM

CLIENT: Municipality of Enschede
ARCHITECTS: Brookes Stacey Randall (London) and IAA (Enschede)
ENGINEERS: IAA (Enschede) and Price & Myers (London)

REFERENCE

Stacey, M. (2001). Presentation at Passenger Terminal Expo, Cannes.

6.6.7 Rotterdam Central Station, The Netherlands

Interchange – national railway station and multi-line metro station with trams and buses adjacent.

A vision of a new style of urban or metropolitan rail station is provided by this project master plan, drawn up by Will Alsop.

This station is the focus of transport to and from this city, being the stop for the European high-speed train as well as the new Randstad Rail route. It is estimated that in 10 years' time, 70–80 million people will pass through each year.

The square in front of the station has been dubbed 'Suicide Square' because, according to Will Alsop, a passenger emerging from the station risks being run over first by a bike, then a taxi, then a tram, then a bus, then a car and again by a bike.

The new transport intersection will be achieved by lowering the existing pedestrian tunnel underneath the rail tracks for passengers changing trains and creating a 'balcony' for pedestrians to move between city and station.

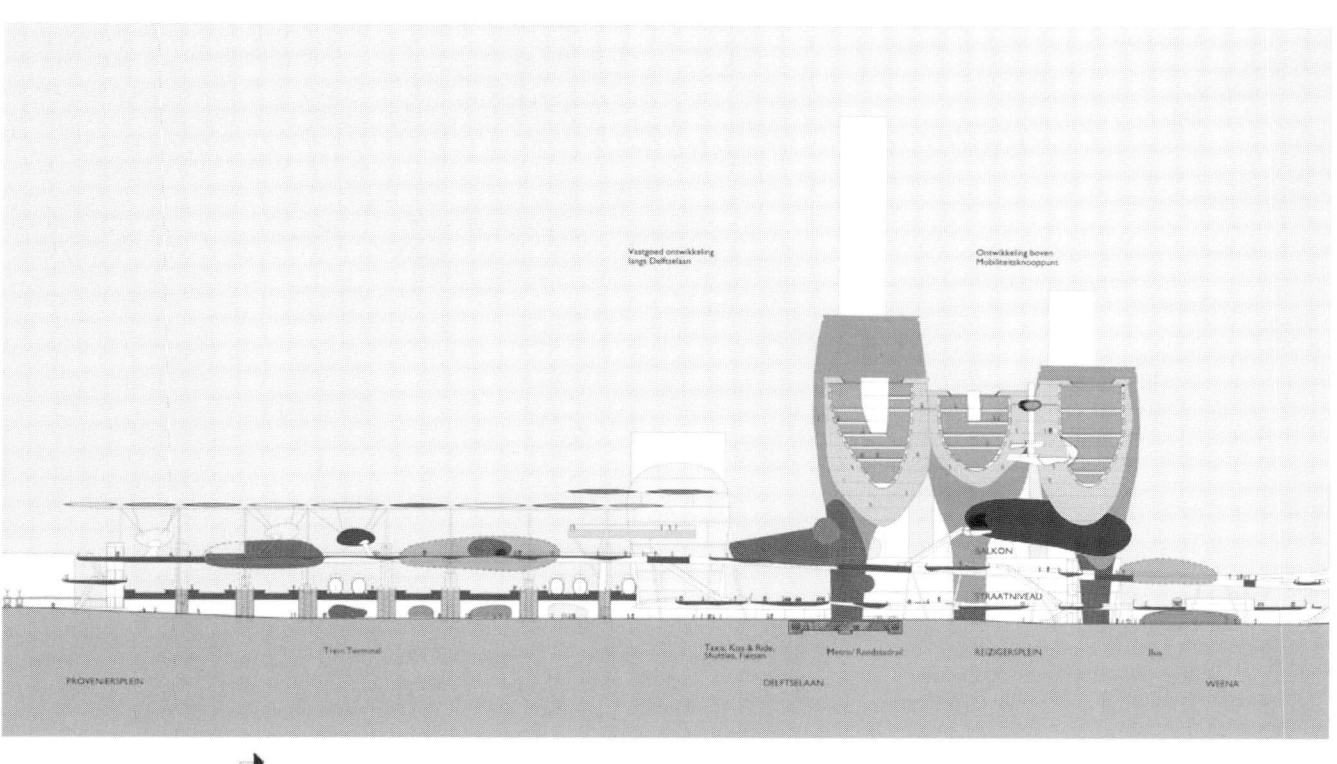

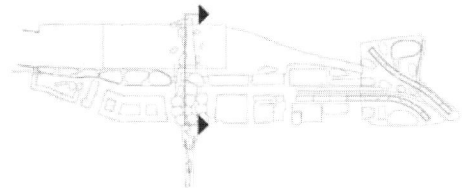

6.149 Rotterdam Central Station, The Netherlands: north–south section (Courtesy of Alsop Architects).

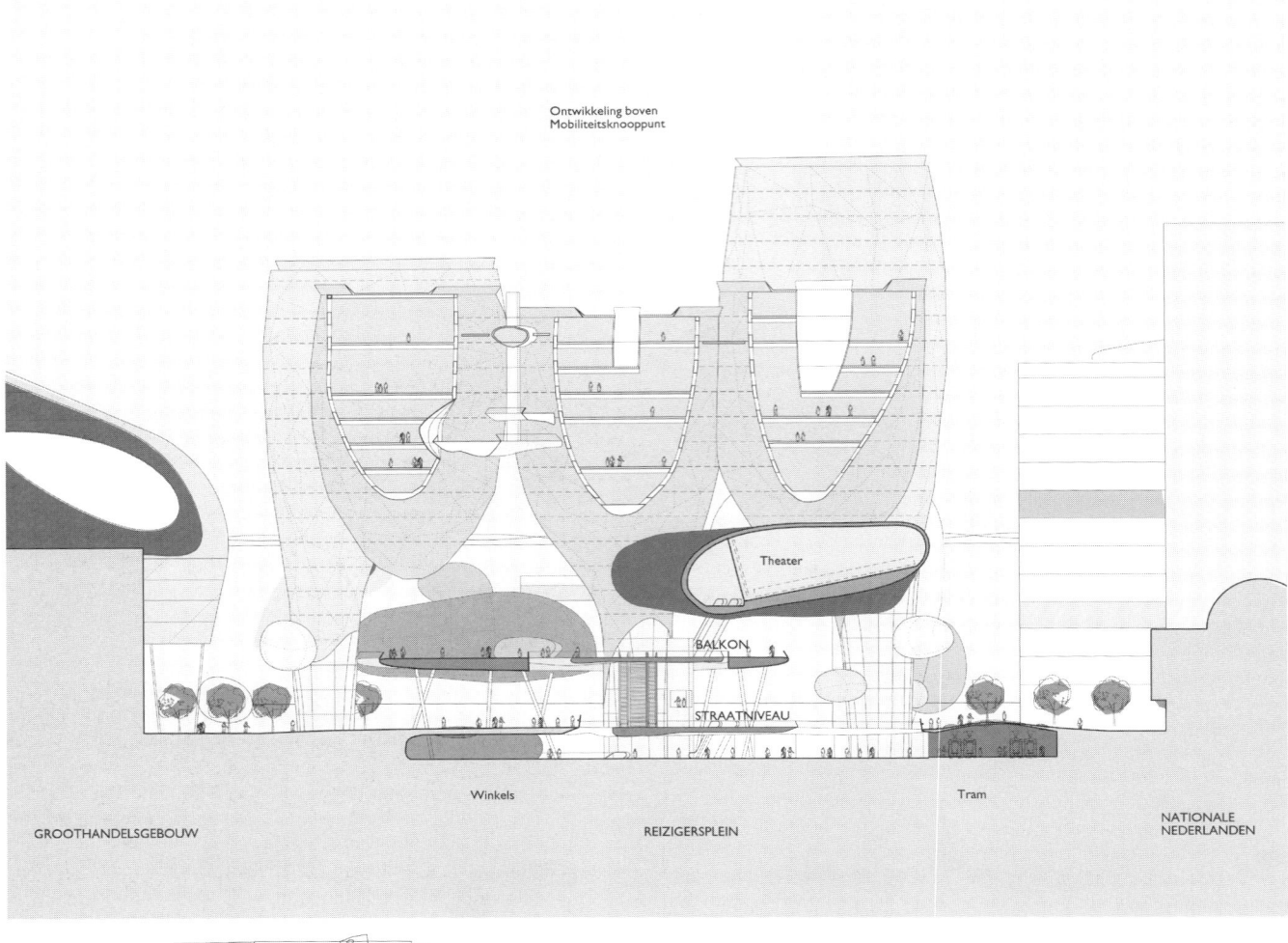

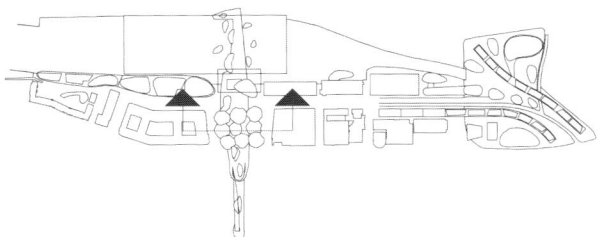

Transport terminals and modal interchanges

6.150 East–west section (Courtesy of Alsop Architects).

Transport terminals and modal interchanges

▲ 6.151 Exploded isometric (Courtesy of Alsop Architects).

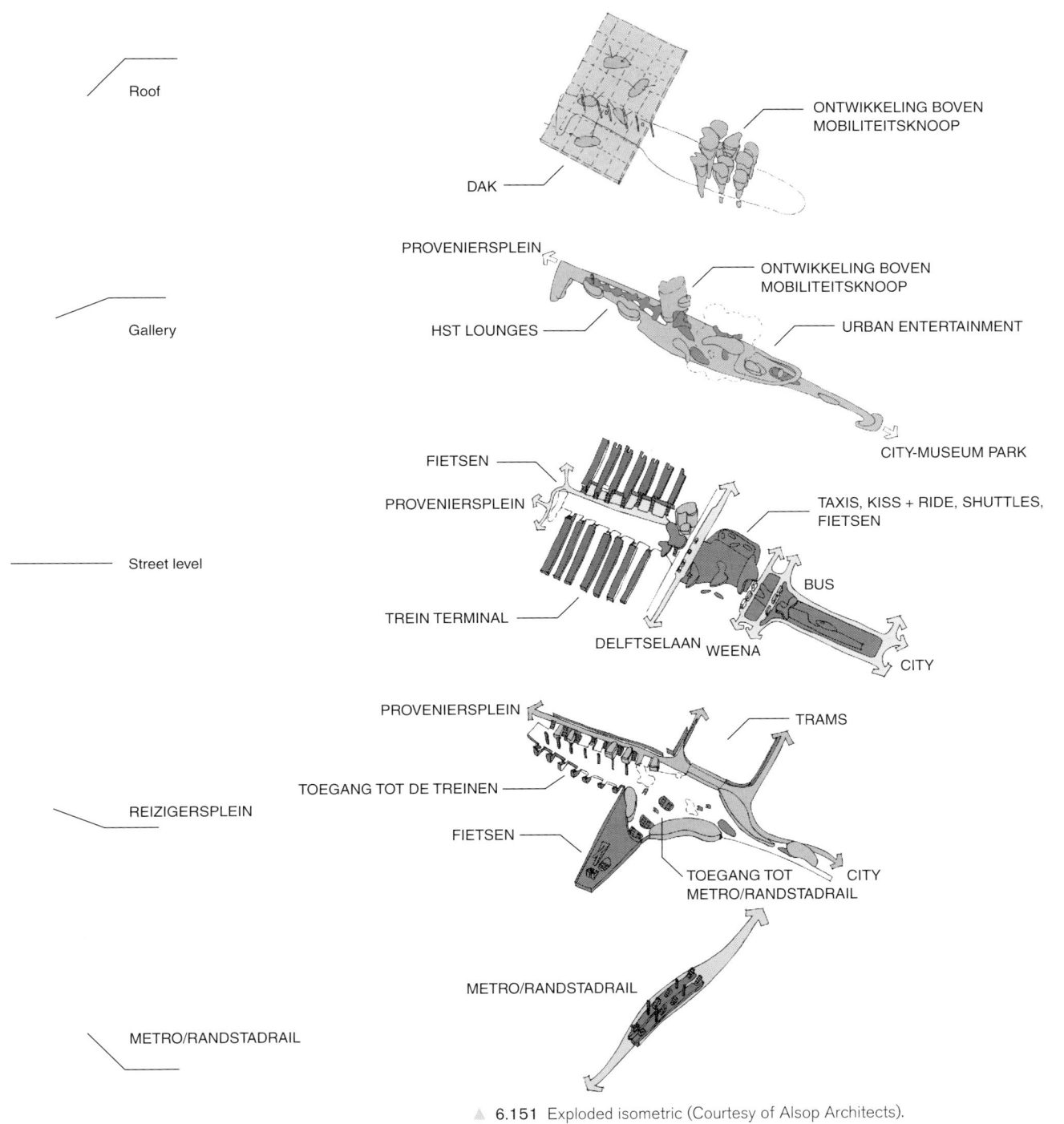

REFERENCE

ARCHIS (2000). Mobility edition, interview with W. Alsop, November.

Multiple railway station/bus and coach/car interchanges: contiguous **151**

6.6.8 Perth Station, Western Australia

Interchange – multi-line urban rail station with buses adjacent.

In developing a new suburban railway system, the City of Perth had many advantages: a relatively small population, in a relatively new, low-density city with plenty of interstitial routes possible. The city station has become an interchange point for four lines which converge and diverge there. The train services have coincident arrival and departure times to facilitate interchange for cross-city journeys, and the city centre itself and sectors not fully served by the railways have bus services radiating from the station or a nearby bus station.

Figures 6.153–6.154 show an interchange moment, with first two then four trains embarking. The free shuttle buses circulating round the city centre leave from the bridge immediately above the platforms in the centre of the pictures and the other local bus station is at the far end of the left-hand platform.

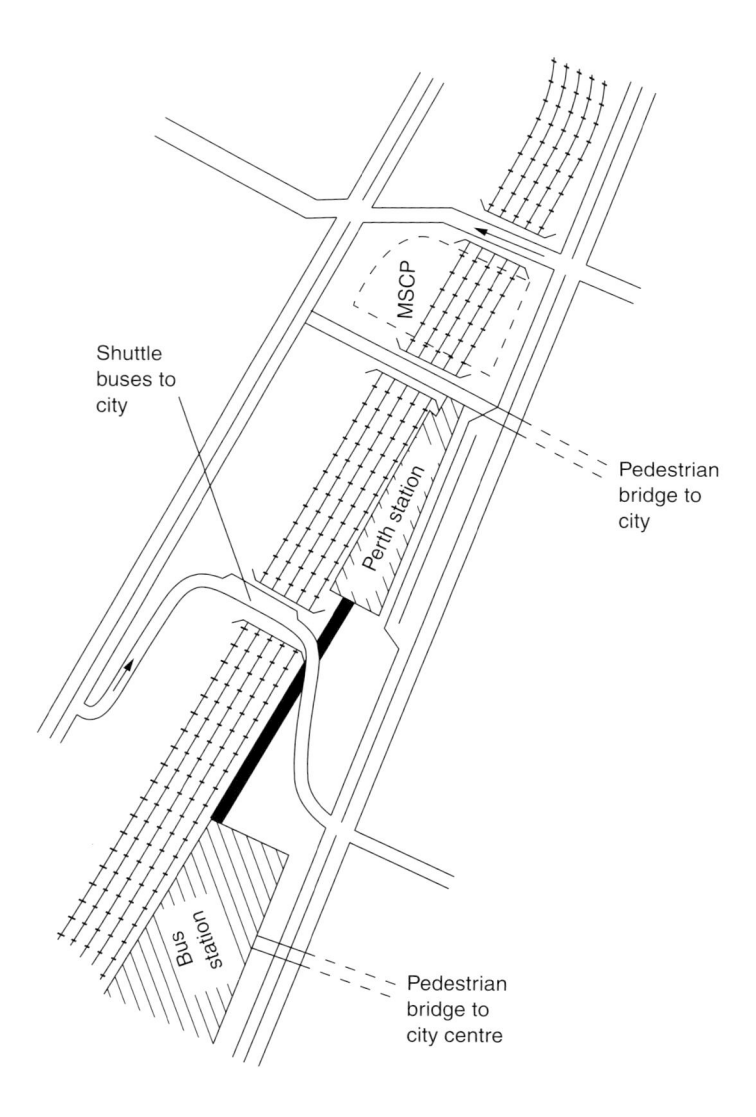

6.152 Perth Station, Western Australia: plan.

Transport terminals and modal interchanges

6.153 View of station with two trains.

6.154 View of station with four trains.

Multiple railway station/bus and coach/car interchanges: contiguous **153**

6.7 Ship and ferry terminals

6.7.1 Southampton Mayflower Terminal, UK

Interchange – cruise liners with road traffic.

A cruise liner arrives or departs at this terminal every other day and each time about 1000 passengers and 4000 items of luggage need to be embarked or disembarked.

New passengers arrive by taxi or coach and need to deposit their bulky luggage already labelled by cabin for security clearance, check in and clear personal security. From a departures lounge they make their way to the gangway to the vessel. Disembarking passengers need to descend from the gangway from the vessel to ground level to reclaim their baggage and clear customs. Neither flow of people is sufficiently frequent or concentrated to justify airport passenger standards, but the sequence needs to be clear and convenient.

The solution adopted at the Mayflower Terminal has been to adapt the existing quayside Shed 106 by dividing it lengthways into a baggage hall and covered pick-up and set-down roads. From the set-down area baggage is passed through straight to the baggage hall and, after check-in and security clearance, passengers enter the departures lounge and are taken by escalator to an elevated walkway on the quayside to board the ship by means of airport-style loading bridges. Disembarking passengers use the same loading bridges and walkway and descend by escalator to the floor of Shed 106 for baggage reclaim and customs.

TEAM

CLIENT: P&O
ARCHITECTS: The Manser Practice

REFERENCE

Baillieu, A. (2003). Cruise control. *RIBA Journal*, July.

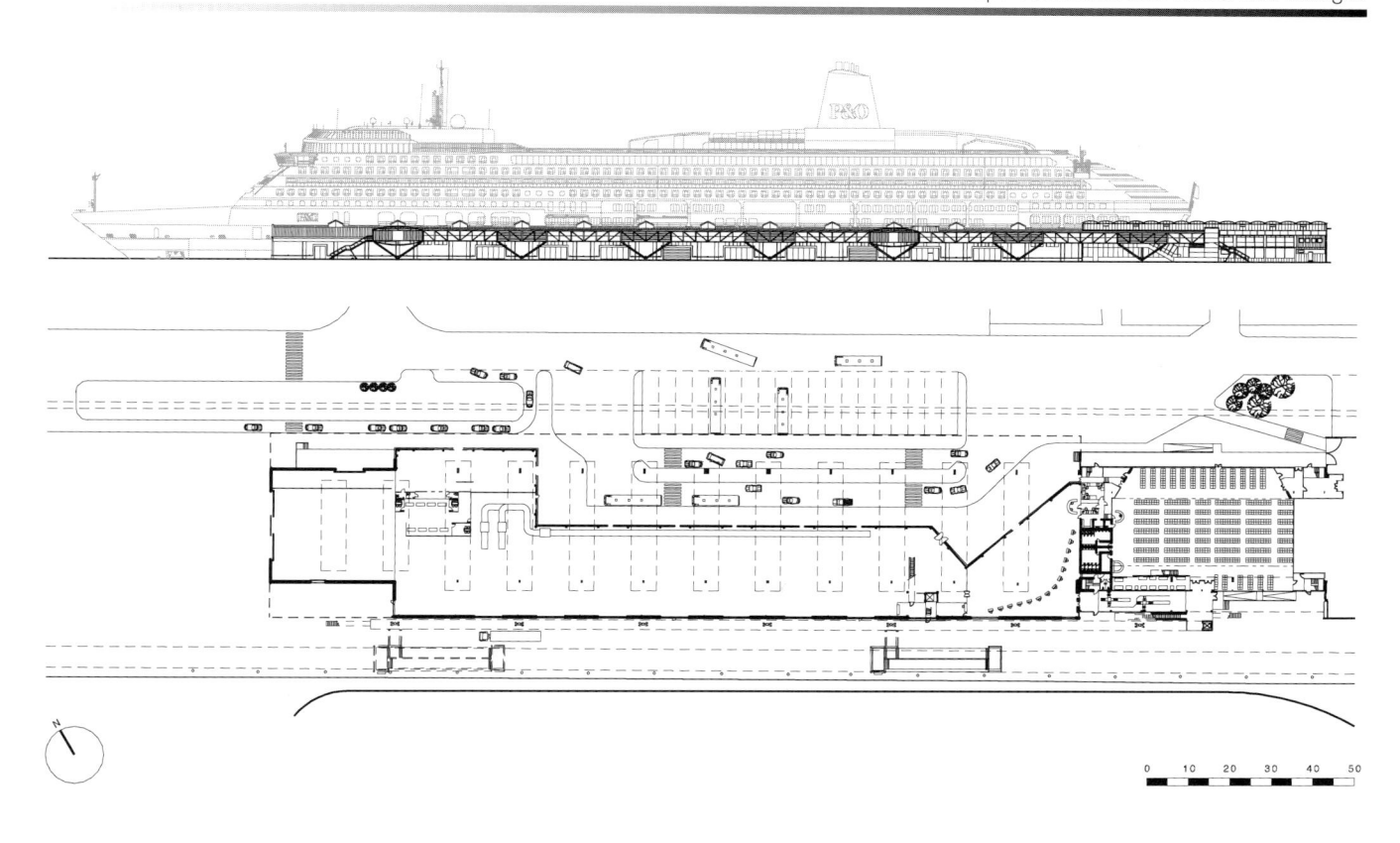

Transport terminals and modal interchanges

MAYFLOWER CRUISE LINER TERMINAL — THE MANSER PRACTICE ARCHITECTS

▲ 6.155 Southampton Mayflower Terminal, UK: long elevation and plan (Courtesy of The Manser Practice).

MAYFLOWER CRUISE LINER TERMINAL — THE MANSER PRACTICE ARCHITECTS

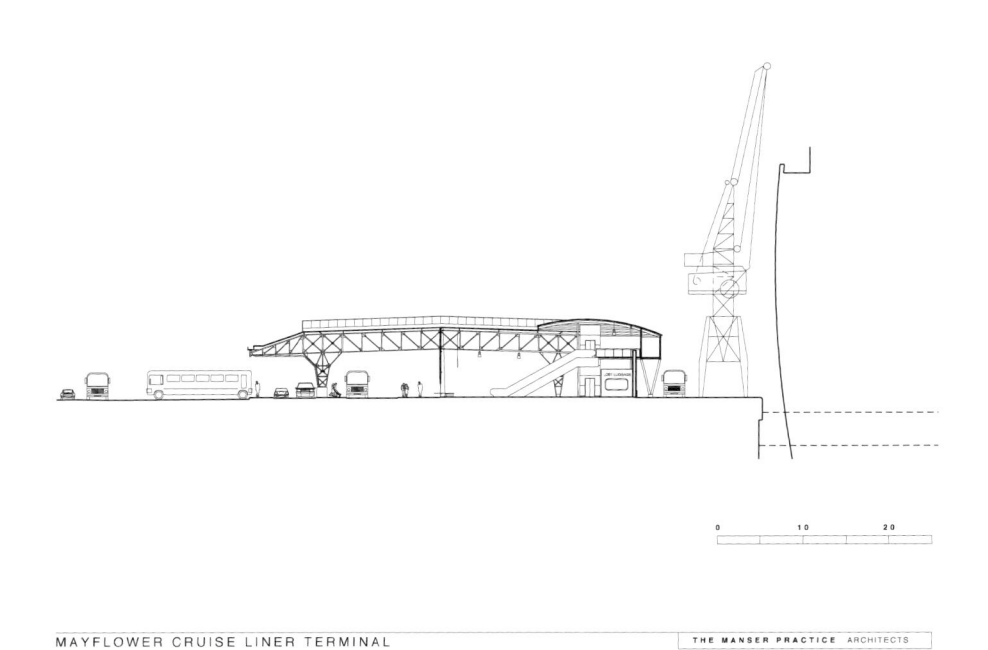

▲ 6.156 Cross-section (Courtesy of The Manser Practice).

Ship and ferry terminals 155

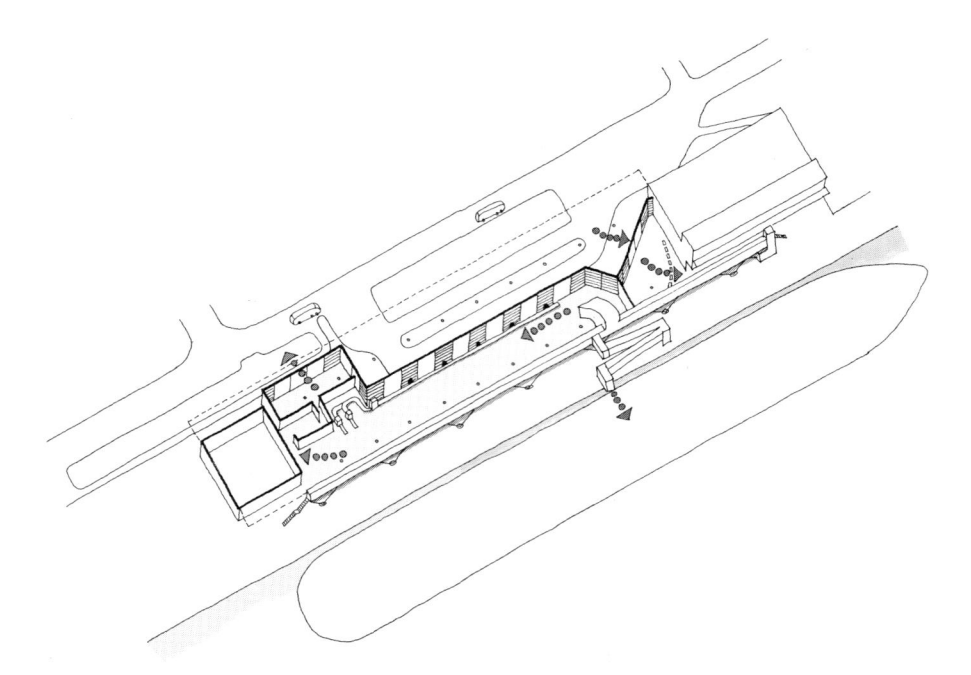

6.157 Southampton Mayflower isometric (Courtesy of The Manser Practice).

Transport terminals and modal interchanges

6.7.2 Yokohama port terminal, Japan

Interchange – cruise liners with road traffic.

To quote from architects Foreign Office Architects Ltd of London:

> The brief asked for the articulation of a passenger cruise terminal and a mix of civic facilities for the use of citizens in one building. The site had a pivotal role along the city's waterfront that, if declared a public space, would present Yokohama City with a continuous structure of open public spaces along the waterfront.

The undulating roof and plaza 'aspires to eliminate the linear structure characteristic of piers and the directionality of circulation'.

TEAM

CLIENT: The City of Yokohama Port & Harbour Bureau Construction Department
ARCHITECTS: Foreign Office Architects Ltd with local architect GKK

REFERENCE

Dawson, S. (2002). Roller-coaster construction. *Architects Journal*, Metalworks supplement, 28 March.

6.158 Yokohama port terminal, Japan: aerial photo with city in background (Courtesy of Foreign Office Architects).

6.7.3 Sydney Overseas Passenger Terminal, Australia

Interchange – cruise liners with road traffic.

This terminal replaces a building dating from 1960 and catering for ocean liners bringing immigrants, a function which declined within 10 years, and the building subsequently served sporadic cruise ships. The new building, which retains some of the structure of the old, has a customs hall at the upper level to handle 2000 passengers per hour and such other facilities as required for the reprovisioning of cruise liners. Cafés and restaurants ensure that the building, in such a dramatic setting on Sydney Cove opposite the famous Opera House, is used as much by diners and dancers as by passengers, workers and ships.

TEAM

CLIENT:	NSW Public Works Department and the Maritime Services Board
ARCHITECTS:	Lawrence Nield & Partners Australia Pty Ltd in association with PWD Government Architects Branch
ENGINEERS:	Ove Arup & Partners

▲ 6.159 Sydney Overseas Passenger Terminal, Australia: view of OPT across Sydney Cove.

Transport terminals and modal interchanges

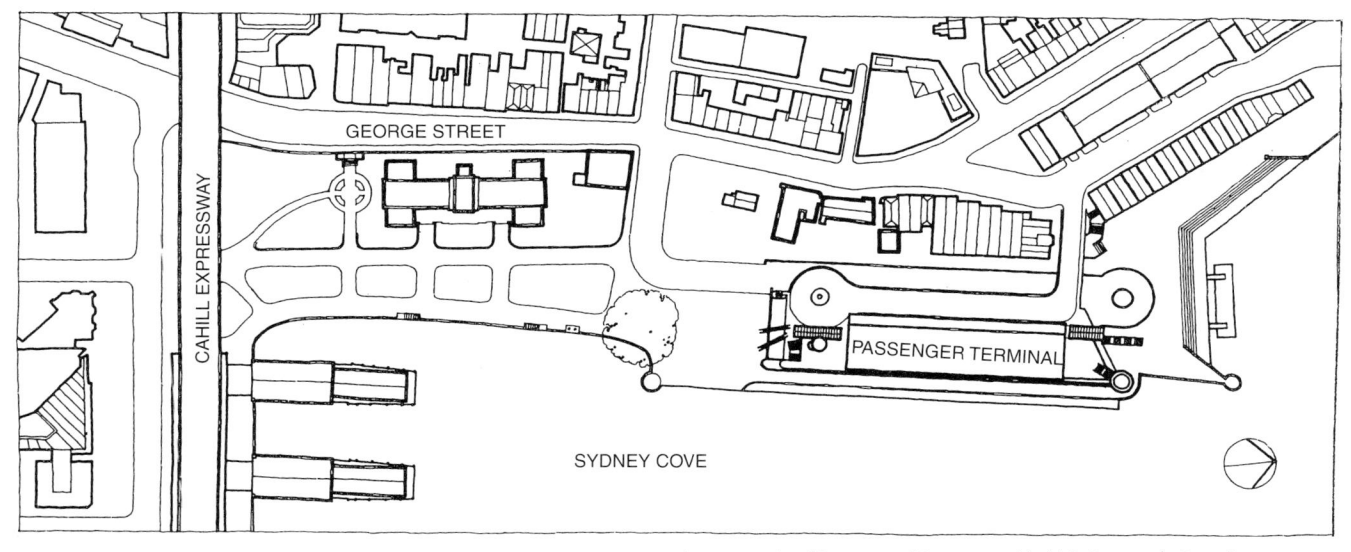

6.160 Location plan (Courtesy of Lawrence Nield & Partners). See also Figure 6.123.

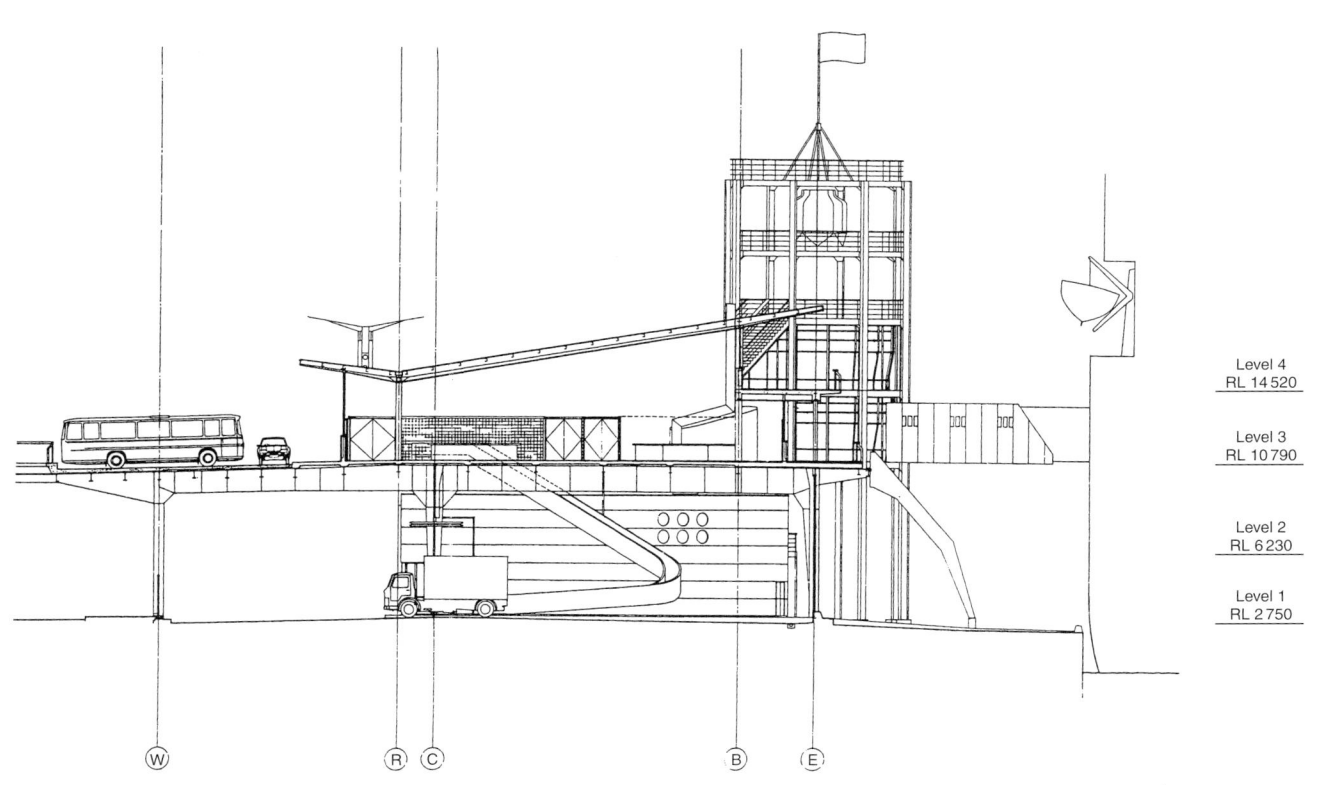

6.161 Cross-section (Courtesy of Lawrence Nield & Partners).

Ship and ferry terminals 159

Transport terminals and modal interchanges

▲ 6.162 Close-up photo at water level (Courtesy of Lawrence Nield & Partners).

160 Ship and ferry terminals

▲ 6.163 Tower at north end acting as urban pivot (Courtesy of Lawrence Nield & Partners).

▲ 6.164 View from south across Rocks Place, a new public space (Courtesy of Lawrence Nield & Partners).

7 Common standards and requirements

This part addresses the aspects of functional requirements that are common to all terminals where passengers board aeroplanes, buses and coaches or railway trains or transfer between them:

- Space standards.
- Security.
- Border controls.
- General legislation.
- Needs of passengers with reduced mobility.
- Commercial opportunities.
- Terminal operator's requirements.
- Transport operator's requirements.
- Car parking.

Note that for practical purposes the consideration of baggage systems is limited to the airport interface, discussed in Chapter 10, justified by the prevailing distinction explained in Chapter 3.1.1.

7.1 Space standards

One man's congestion is another's profit: space standards are variable and subjective. The objective solution is to quote from the concept of Level of Service. The application of this is common to all terminals and interchanges, and the differences arise, for example, from the amounts of baggage involved. Table 7.1 shows levels of service related to unit space standards in different types of space. For many passengers the criterion by which terminals such as airports are judged is the walking distance between one mode of transport and another. Although there is an inevitability about the length of a railway station platform or an airport pier, design can mitigate the strain of walking distance by providing passenger conveyors.

Table 7.1 takes standards for the airport sector and generalises them.
The most comprehensive review of the space that people need to walk, queue, crowd, wait, access buses, trains, planes, lifts and escalators is found in *Pedestrian Planning and Design*, by J. J. Fruin.

7.2 Security

In the case of air travel in particular, but also in principle for long-distance rail and sea travel, the checking of passengers and their possessions requires the installation and manning of suitable equipment and the strategic location of the check point in order to ensure both that no passengers evade or avoid the checking procedure and that the procedure is carried out in the most efficient manner.

The events of 11 September 2001 have given new impetus to security for travellers and transport installations. Specialist equipment for screening baggage is outside the scope of this book.

Table 7.1 Levels of service and space standards (m² per person)

	Levels of service				
	A	B	C	D	E
Areas with trolleys					
Baggage check-in queues					
A Few trolleys and little baggage	1.7	1.4	1.2	1.1	0.9
B Few trolleys and 1–2 baggage items/person	1.8	1.5	1.3	1.2	1.1
C Most people with trolleys	2.3	1.9	1.7	1.6	1.5
D Most people with trolleys + lots of baggage	2.6	2.3	2.0	1.9	1.8
Waiting and circulating with trolleys			2.3*		
Dense, waiting to reclaim baggage, 40% trolleys	2.6	2.0	1.7	1.3	1.0
Areas without trolleys					
Waiting and circulating without trolleys (e.g. airside)			1.5**		
Queuing (e.g. Immigration) without trolleys	1.4	1.2	1.0	0.8	0.6

* Speed of movement possible at level C space standard: 0.9 m/s.
** Speed of movement possible at level C space standard: 1.3 m/s.
Level A: excellent service, free flow, no delay, direct routes, excellent level of comfort.
Level B: high level of service, condition of stable flow, high level of comfort.
Level C: good level of service, condition of stable flow, provides acceptable throughput, related subsystems in balance.
Level D: adequate level of service, condition of unstable flow, delays for passengers, condition acceptable for short periods of time.
Level E: unacceptable level of service, condition of unstable flow, subsystems not in balance, represents limiting capacity of the system.
Source: adapted from IATA *Airport Development Reference Manual* (2004 edition).

7.3 Border controls

Quite apart from security considerations, many terminals occur at national borders and therefore are the point of entry to or exit from sovereign areas. Accordingly, customs and immigration controls need to be conducted.

7.4 Building design legislation

Places of assembly of large numbers of people require special consideration of means of escape in case of fire, as well as the normal controls on the standard of building construction.

7.5 Needs of passengers with reduced mobility

Irrespective of national legislation, such as the Disability Discrimination Act in Britain, the IATA *Airport Development Reference Manual* (2004 edition) recommends a level of service to be achieved in airport terminals, a discipline which can be applied to all passenger terminals. It advocates the drawing up of Disabled Access Assessment Plans, with stars recorded for various measures, culminating in a total score for a building and the award of overall gold, silver and bronze star ratings.

7.6 Commercial opportunities

Wherever large numbers of people assemble, and particularly wait, they need catering and business facilities. If they have money to spend there will be any number of shopping opportunities.

7.7 Terminal operator's requirements

The owner and/or the operator of the terminal will be out to make the maximum return on investment and this will probably involve collecting revenue from the transport operator and the commercial concessionaire rather than the passengers or the public.

In a move designed to improve the image of the airport operator, as well as improve the lot of the passenger, the Airports Council International in 2001 launched 11 commitments as a challenge to its members:

1. Services for those with reduced mobility.
2. Passenger information on legal rights.
3. Assistance during periods of significant delay or disruption.
4. Improved airport access and ground transportation.
5. Provision of infrastructure for check-in, baggage and security.
6. Systematic maintenance of equipment.
7. Trolley availability and quality.
8. Wayfinding and information desks.
9. Cleanliness.
10. Response to comments and complaints.
11. Annual reports on passenger satisfaction and performance indicators.

7.8 Transport operator's requirements

On the other hand, the transport operator will want to get the passengers through the building as quickly as possible. Functional performance is paramount and related to speed, and the requirement for speed and efficiency is accentuated by the transfer facility. For example, the transfer time between connecting flights at airports is being reduced again and again to provide a 'hub', and as different transport systems are integrated interchange times between, for example, train and plane need to be improved.

7.9 Car parking

This book does not analyse and review car parking demands and solutions at airports and railway stations, but examples are apparent throughout Chapter 6. Notable solutions are of two types, arising from chronological patterns of development. Where car parks came before public transport linkages, the car park adjacent to the terminal becomes a prime location for the ground transport interface, as at airports like Manchester and Chicago. Where the design and construction of the interchange is more recent, as at Ashford, car parking structures are an integral part of the interchange.

REFERENCES

Adler, D. (ed.) (1999). In *Metric Handbook*, 2nd edition, Chapter 4, Part 7 – Parking. Architectural Press.

Fruin, J. J. (1971). *Pedestrian Planning and Design*. Metropolitan Association of Urban Designers and Environmental Planners Inc.

International Air Transport Association (2004). *Airport Development Reference Manual*, 9th edition. IATA.

8 Bus and coach interface

A bus station is defined as an area away from the general flow of road vehicles, which gives buses and coaches the freedom of movement to set down and pick up passengers in safety and comfort. Locations are either near shopping centres or other transport terminals, thereby affording the best interchange.

Two particular trends have affected urban bus and coach operations: the use of one-man buses for economy of manpower, and deregulation leading to entry into the marketplace by new companies with new operating methods and, in many cases, minibuses.

8.1 Vehicles

See Figures 8.1–8.7.

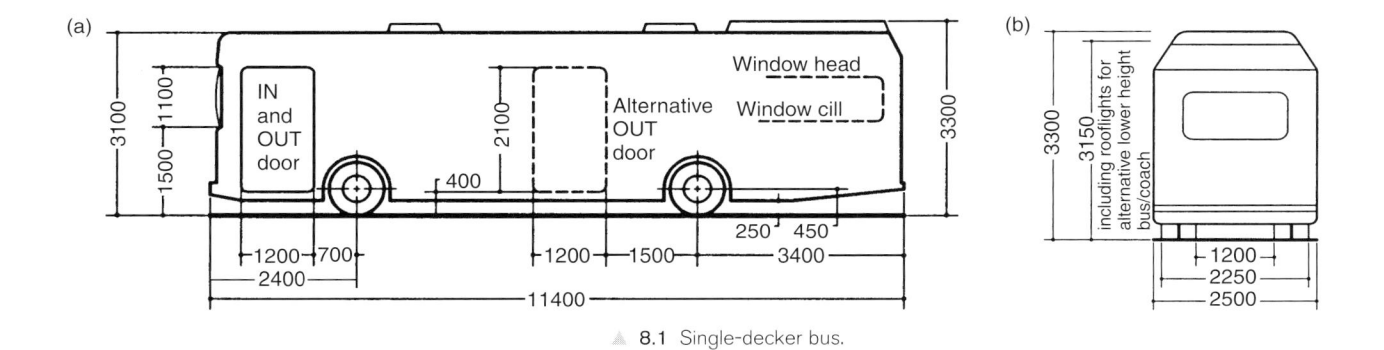

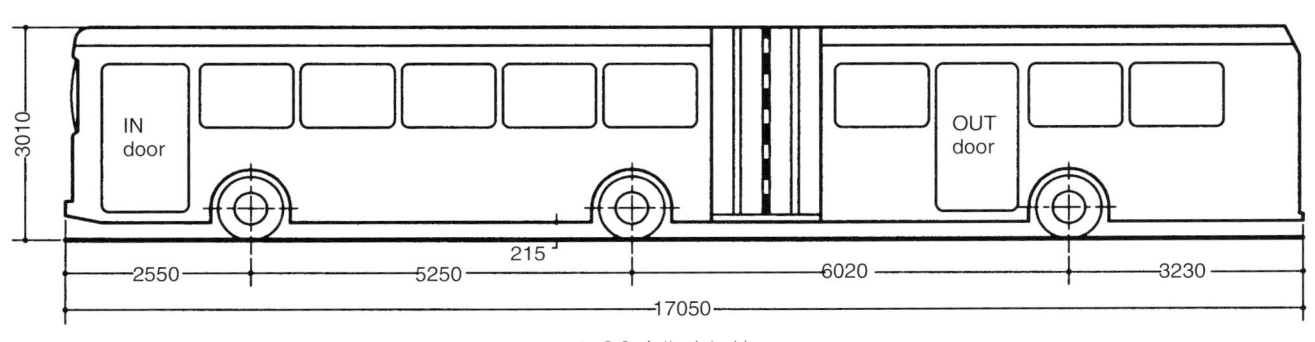

8.1 Single-decker bus.

8.2 Articulated bus.

Bus and coach interface **165**

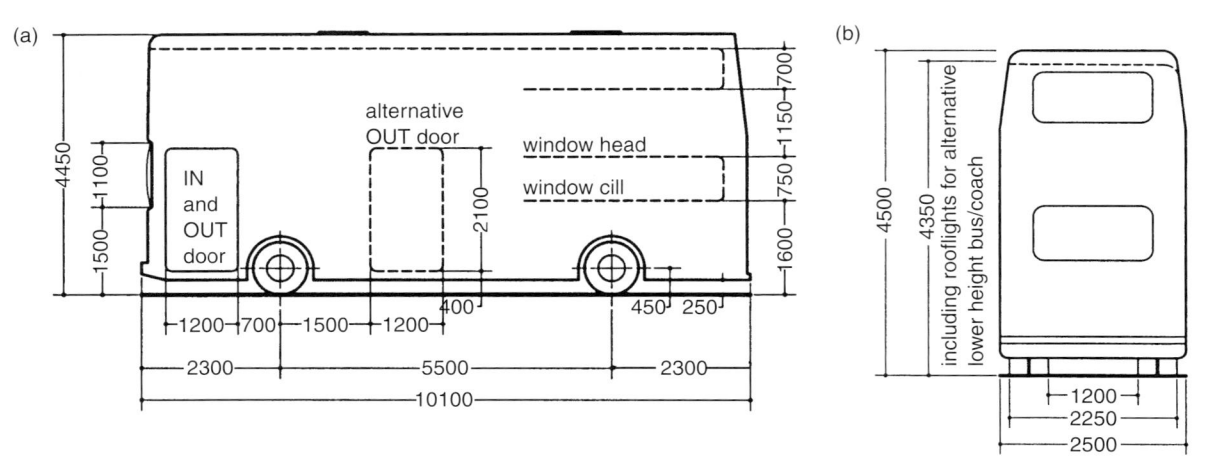

Transport terminals and modal interchanges

(a)

(b)

8.3 Double-decker bus.

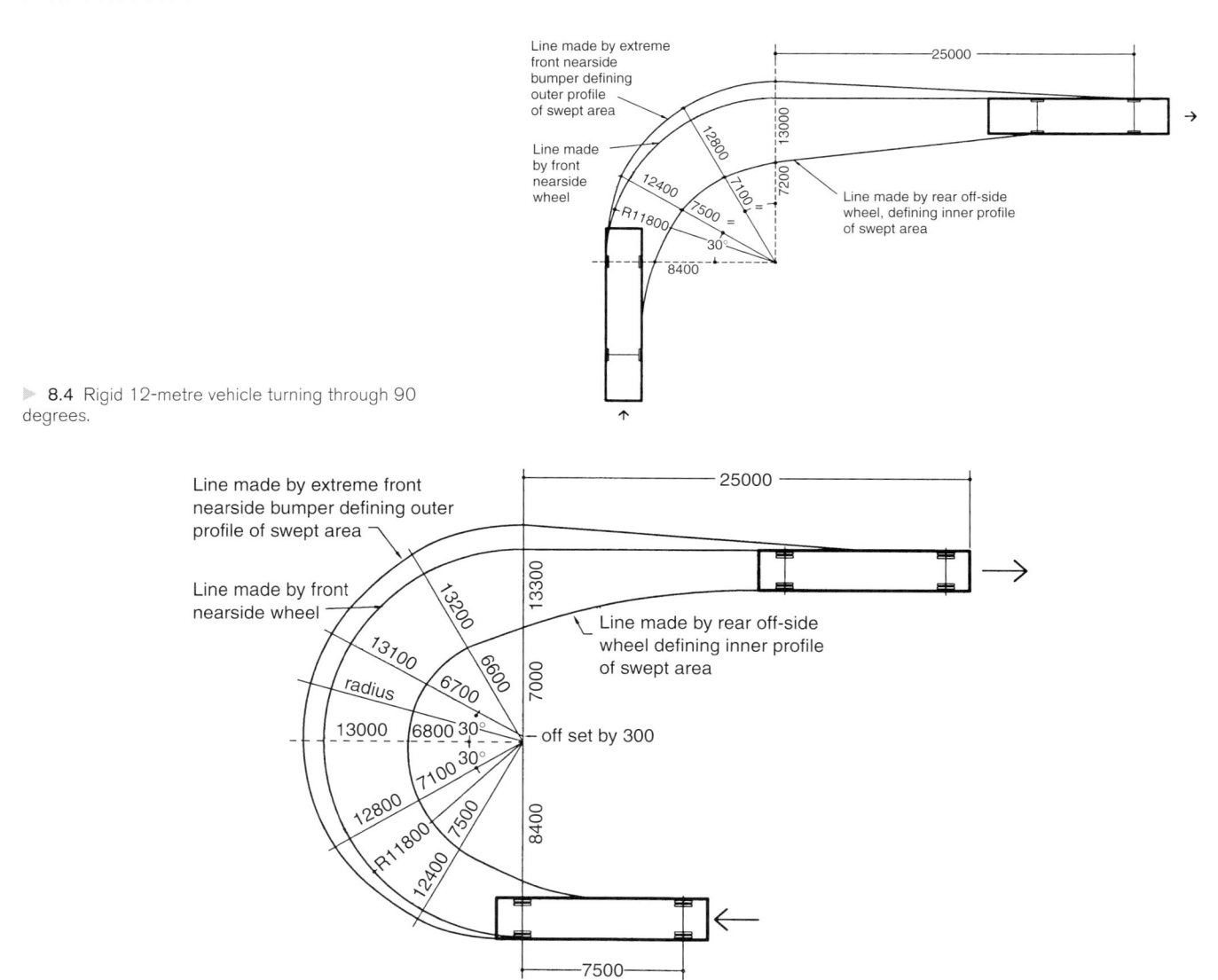

8.4 Rigid 12-metre vehicle turning through 90 degrees.

8.5 Rigid 12-metre vehicle turning through 180 degrees.

166 Bus and coach interface

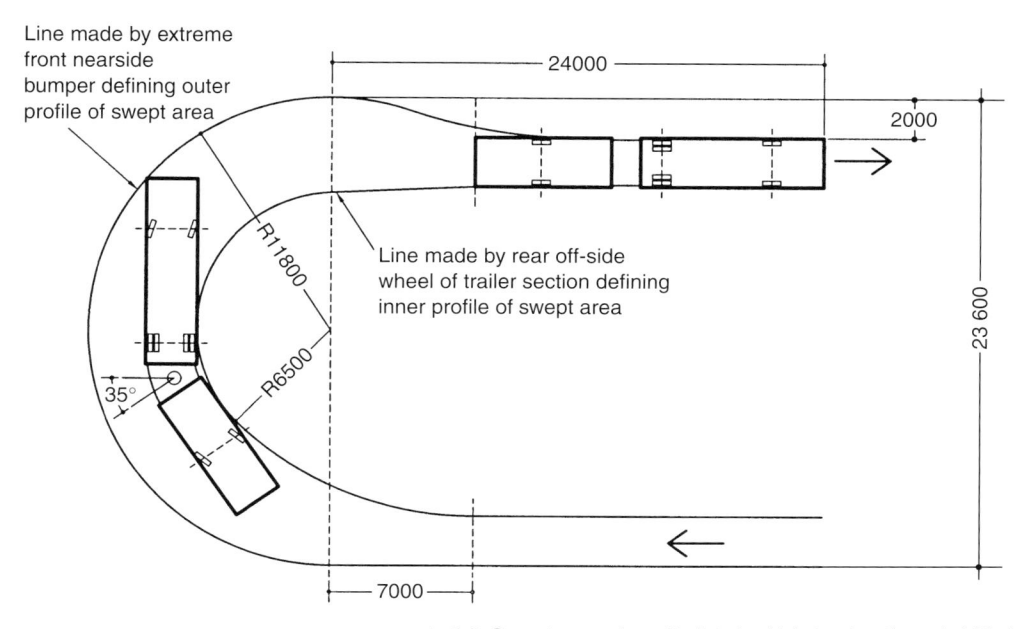

Transport terminals and modal interchanges

8.6 Seventeen-metre articulated vehicle turning through 180 degrees.

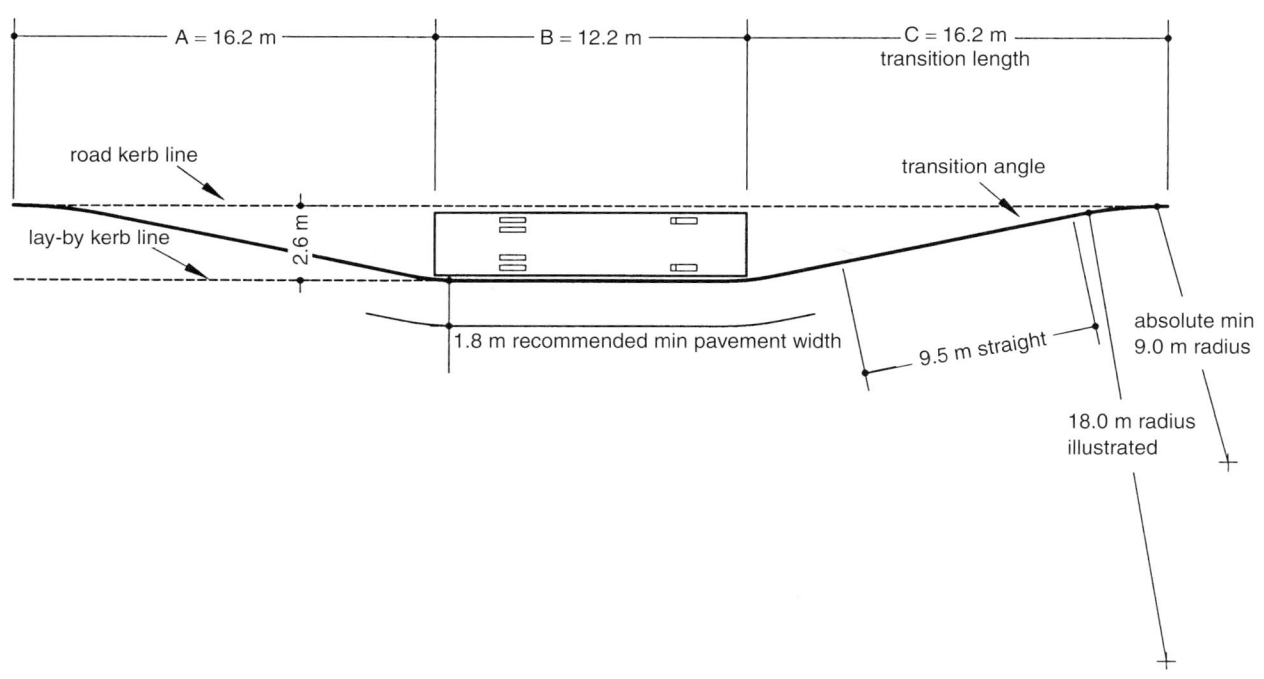

8.7 A lay-by with one bus stop, assuming normal urban speed of approach. The transition length of 16.2 m is the minimum for a 12-m rigid vehicle. Three bus stops is the desirable maximum in a lay-by, the maximum comfortable distance for a pedestrian to walk. Overall length is A + nB + C, where n is the number of buses to be accommodated.

Bus and coach interface **167**

8.2 Factors affecting size of station

Stations will vary in size governed by the following basic points, apart from the obvious physical constraints of the site:

- The number of bays to be incorporated (the term 'bay' is used in connection with stations instead of the term 'bus stop'), determined by the number of bus and coach services to be operated from the station, and by how practical it is, related to the local timetable, to use an individual bay for a variety of service routes.
- The vehicle manoeuvre selected to approach the bays. Three basic types of manoeuvre are used, namely 'shunting', 'drive-through' and 'sawtooth'. The choice of manoeuvre will be influenced by the size and proportions of the site available, the bus operators' present and anticipated needs, and in particular the preference of their staff. Some will accept the sawtooth arrangement while others prefer the drive-through. The area of the site is further added to by the requirement of 'layover'. This is where vehicles having set down their passengers, but which are not required to collect passengers, are parked on the station until needed again. The layout for this should be based on the requirement for parking, but preferably in such a manner that no vehicle is boxed in by another, and of course positioned so as not to interfere with other bus movements. In some cases economy of space can be achieved, again dependent upon local timetables, by using spare bays for layover purposes.
- The facilities to be provided for passengers. Provision for passengers will depend entirely upon anticipated intensity of use and the multi-modal nature of the interchange. If, for example, there are already public toilets, a bus and coach information centre and cafés nearby, then these may not be required on the station concourse. However, waiting room facilities will probably be required, with someone on hand to give information and supervision. In more comprehensive schemes, in addition to a waiting room, a buffet and public toilets, one may plan for kiosks and enquiry, booking, left luggage and lost property offices.
- The facilities to be provided for staff. There will invariably be an inspector or inspectors in a station who, as well as assisting passengers, are primarily concerned with supervising the comings and goings of vehicles, their drivers and conductors. If there is a depot near to the station then most staff facilities will be provided there. However, if the depot is some distance away, it will be necessary to provide canteen and toilets for them on the station site, so that during breaks and between working shifts they do not need to get back to the depot until they return their vehicle for long-term parking. Should the depot be even more remote, it will be necessary to provide all facilities at the station site and only basic amenities at the depot. In this case, as well as the canteen and toilets, a recreation area, locker rooms and 'pay-in' facilities should be provided. The latter is an office area where drivers/conductors check, then hand over monies taken as fares, which in turn are checked and accounted for by clerical staff.
- Facilities for bus maintenance. It will be appreciated that the proper inspection, repair and servicing of buses and coaches is an integral part of a bus operator's responsibility. Normally, such work would be carried out at a local depot, with a repair workshop together with fuelling, washing and garaging facilities. The provision of some or all of these facilities within a station complex is unusual, but by no means unique. For a new

Transport terminals and modal interchanges

(a) Passengers set down only

3 moves to 2   2 moves to 1   1 moves away

13000 min

(b) Bay 1   Bay 2   Bay 3

Clearly defined pedestrian crossing

2000 minimum
6000 minimum

(c) Bay 1 2 3 4 5 6

Angle of pitch

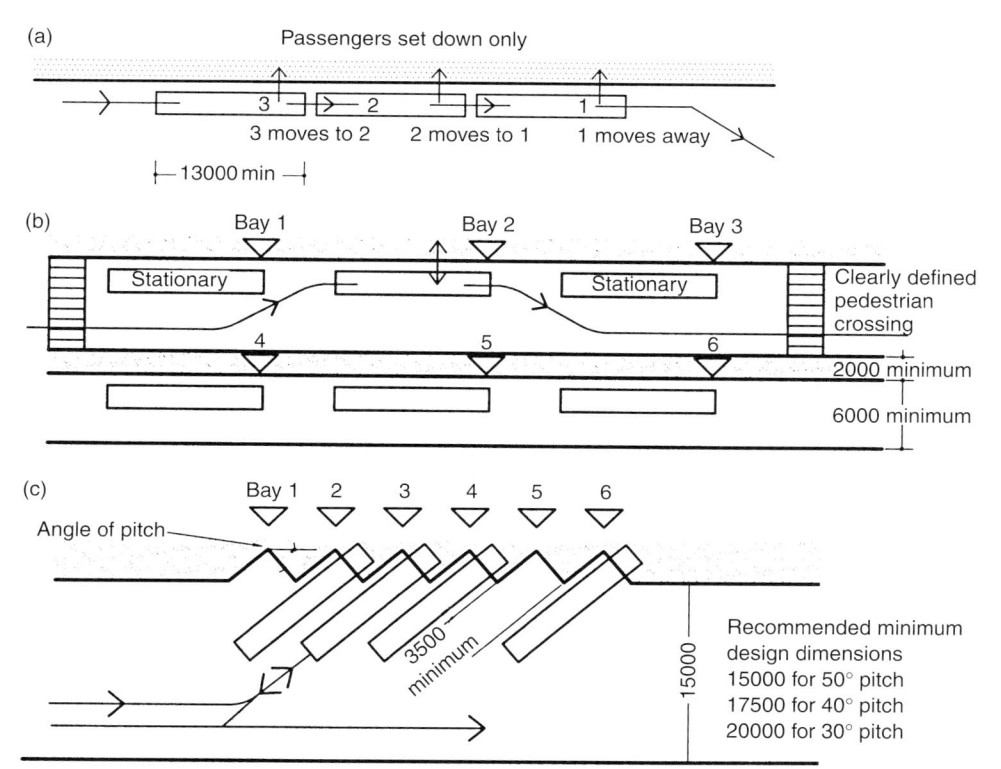

3500 minimum

15000

Recommended minimum design dimensions
15000 for 50° pitch
17500 for 40° pitch
20000 for 30° pitch

▲ 8.8 Vehicle manoeuvres used in approaching parking bays. (a) Shunting is used where a vehicle only sets down passengers on to their concourse before moving away to park or to a bay position for collecting passengers. This manoeuvre avoids waiting to occupy a predetermined bay and effectively reduces journey time.
(b) Drive-through bays are fixed bay positions for setting down and/or collecting passengers. They are in a line, so a vehicle often has to approach the bay between two stationary vehicles. In practice, it is often necessary to have isolated islands for additional bays with the additional conflict of passenger and vehicle circulation.
(c) 'Sawtooth' layouts have fixed bay positions for setting down and/or collecting passengers with the profile of the concourse made into a sawtooth (sometimes referred to as echelon) pattern. In theory, the angle of pitch between the vehicle front and the axis of the concourse can be anything from 1 to 90 degrees. In practice, however, it usually falls between 20 and 50 degrees. The vehicle arrives coming forward and departs going backwards, thus reducing the conflict between passenger and vehicle, but demanding extra care to be taken when reversing out of the bays.

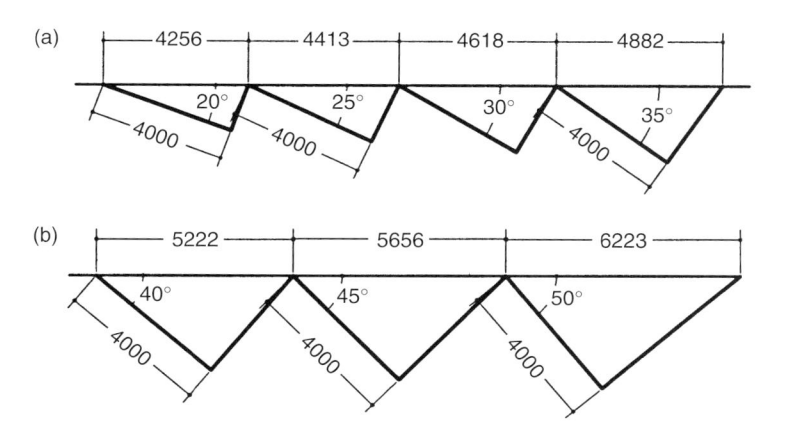

(a) 4256 — 4413 — 4618 — 4882
20° 25° 30° 35°
4000 4000 4000 4000

(b) 5222 — 5656 — 6223
40° 45° 50°
4000 4000 4000

▶ 8.9 As the angle of pitch in sawtooth bays increases, so does the distance between each bay.

Bus and coach interface

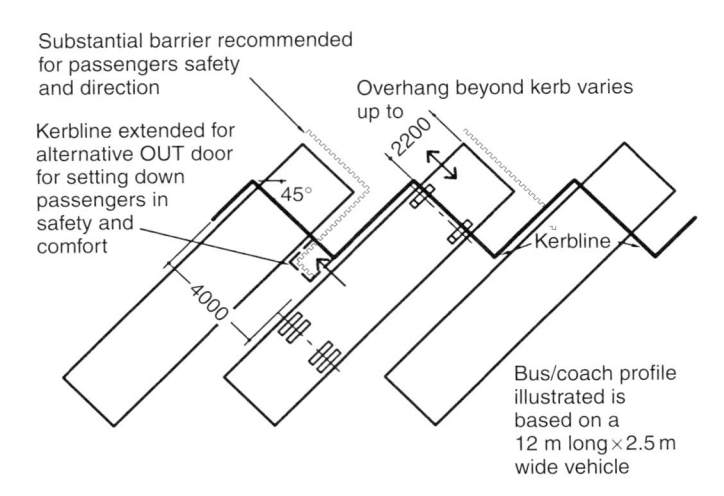

8.10 Passenger safety and control are particularly important when detailing sawtooth bays.

town bus station or for a station in a traffic congested township, where it will be difficult and time-consuming to drive to and from the station and depot, the inclusion of at least a workshop would be advantageous.

Having established the accommodation to be provided on the station site, the problem is then to combine them in a well-planned arrangement.

9 Rail interface

9.1 Heavy rail systems

This part covers platform and related bridge structure requirements only. In other respects, railway stations have the common components of passenger terminals: concourses, ticket offices and commercial outlets. In fact, with the converging standards referred to in Chapters 3 and 7, public spaces will be indistinguishable.

Dimensional standards for railways have progressively converged in Europe, since the days of 'the battle of the gauges'. However, the standardisation of the wheel gauge has not been matched by the loading gauge, and mainland Europe has built coaches and particularly freight vehicles to larger cross-sections. The advent of the Channel Tunnel in 1994 has highlighted the two principal standards for all-purpose stock, while at the same time setting new and quite different standards for dedicated railway stock. The tunnel has been designed to accommodate 800-metre-long trains of 5.6-metre-high wagons, but the conventional coaches in these trains are nevertheless built to the British standard so as to fit under British, and therefore all, bridge structures.

Figures relate to general European and British platforms and bridge structures in section. Platform lengths can vary considerably, but 250 metres is common for main-line stations and 400 metres is the exceptional length of platforms for the London–Paris and London–Brussels trains styled Eurostar.

Note that clearance dimensions are valid for straight and level track only. Due allowance must be made for the effects of horizontal and vertical curvature, including superelevation. Note that the UK DoT standard states that, to permit some flexibility in the design of overhead equipment, the minimum dimension between rail level and the underside of structures should be increased, preferably to 4780 mm, or more, if this can be achieved with reasonable economy.

9.1 Cross-section: controlling dimensions for railway structures, European (Berne gauge) standard.

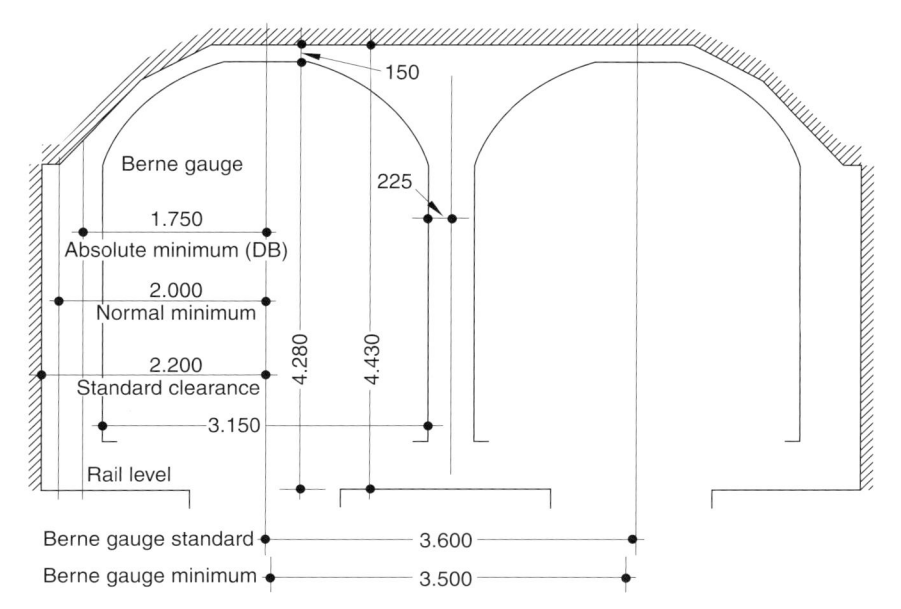

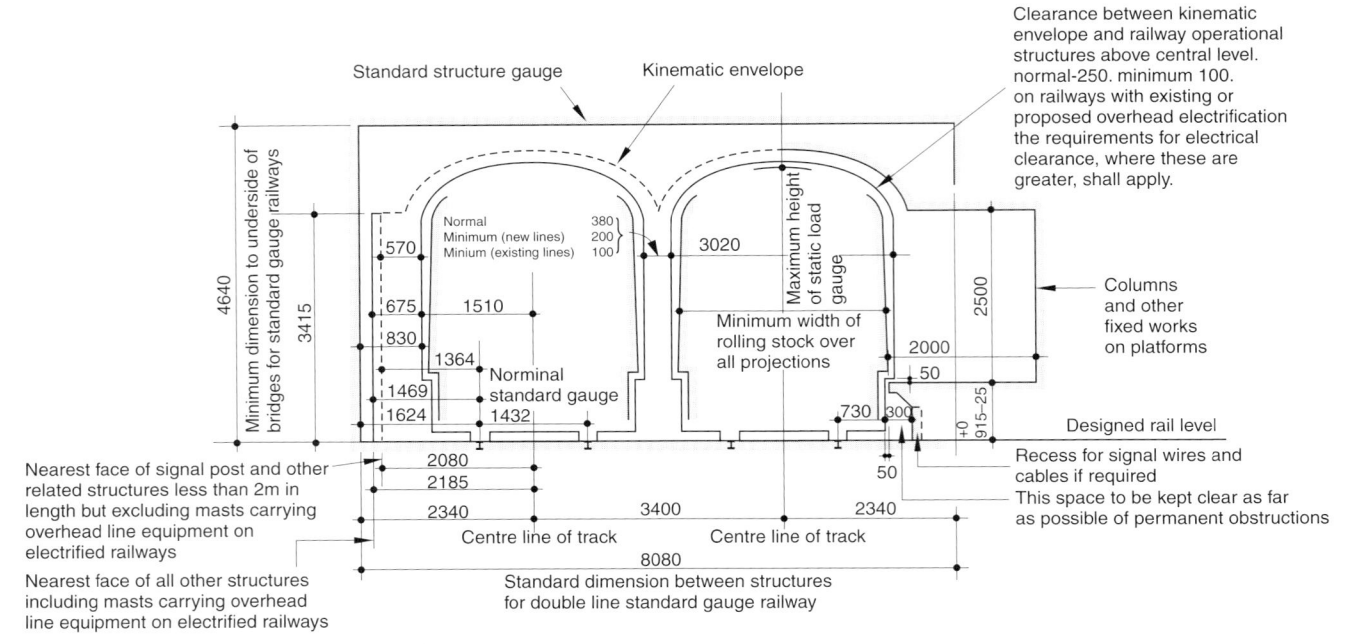

Transport terminals and modal interchanges

▲ 9.2 Cross-section: controlling dimensions for railway structures, British standard.

REFERENCE

Department of Transport (1977). *Railway Construction and Operations Requirements, Structural and Electrical Clearances.* HMSO.

9.2 Light rail systems

Many proprietary systems are installed and feature in many of the 36 interchanges in Chapter 6. One example is dimensioned as shown in Figure 9.3.

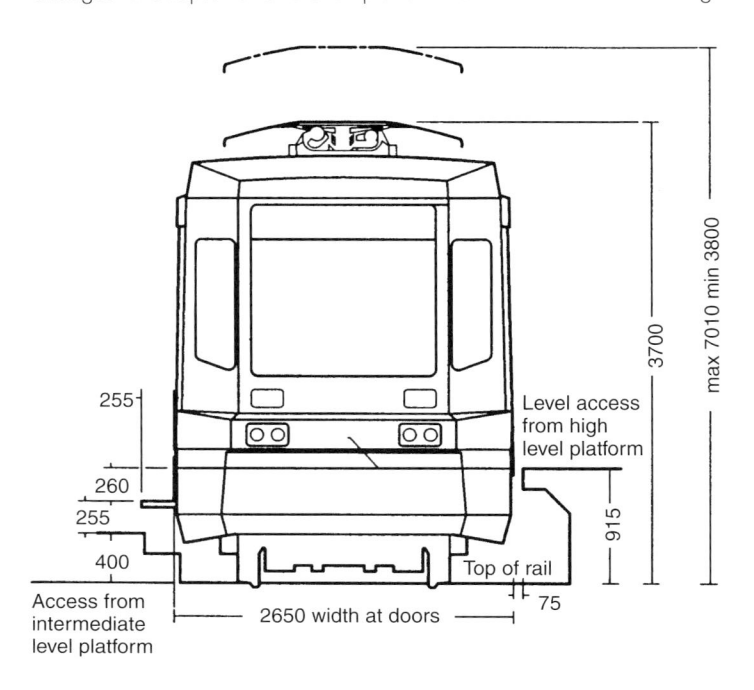

▶ 9.3 Manchester Metrolink: a typical modern tramway system. Frontal view of a car, showing level access for wheelchairs from a high-level platform.

172 Rail interface

10 Airport interface

This part covers planning, landside and airside factors relating to the airport terminal. This has been an established building type for only 70 years, since the time when London's airport was at Croydon for example, but over that time a proliferation of building forms has evolved. All have been responsive to the needs of the moment, but the speed of development of air travel has meant that buildings have rapidly become obsolete and either needed replacement on new sites or internal reconstruction.

A notable early example is the original terminal at Gatwick Airport (Chapter 2), which offered passengers a direct and sheltered route from railway to terminal and from terminal to aeroplane, and was therefore in 1936 one of the first true interchange facilities.

10.1 Airport terminal planning

There are two major influences on airport size and therefore airport terminal size: population demand and airline traffic scheduling. Other factors and forms are listed and described in this section.

Every world metropolis and population centre has by now a giant airport in its vicinity and most cities have airports appropriate to their local needs. Either because the numbers of passengers, flights and choices of destination have increased to a certain level or because of its 'crossroads' location, a particular airport and its one or more terminals can take on a secondary growth pattern. Traffic attracts more traffic, since a wide range of airlines and destinations in turn attracts passengers from a larger area, possibly away from what would otherwise be their nearest airport, and also attracts airlines to feed connecting flights. Ultimately, high volumes of traffic attract airlines to use their routes and facilities to the maximum by creating 'hubs', junctions for radiating routes with convenient flight-changing or transfer facilities for passengers.

10.2 Airport terminal capacity and size

Passengers per hour and passengers per year: these two key factors in airport terminal design are related by traffic distribution. A peak concentration at certain hours of the day will produce a high hourly demand in relation to annual traffic. A constant daily traffic level will produce a high annual rate in relation to the hourly demand.

It has been common for terminal design criteria to be related to the hourly capacity or the number of passengers due to be handled in the thirtieth busy hour of scheduled use. The term SBR or Standard Busy Rate is used. This means that in the 29 hours in the year in which demand is greatest, the facilities will not match the requirement, but it ensures reasonable standards and economy.

Other factors are as follows:

- *Aircraft movements*. Number of arrivals and departures per hour, aircraft sizes, number of stands for each size or range of sizes, passenger load factors.
- *Baggage quantities*. Number of pieces per passenger, by class of travel and traffic (international/domestic).
- *Visitors*. Number of accompanying visitors with departing and arriving passengers by class of traffic (international/domestic).
- *Employees*. Number and proportion for airport, airline, concessionaire, control authorities, etc. and proportion of males and females.
- *Landside transport.* Number of passengers, visitors and employees arriving by private vehicles (note ratio of owner-drivers) and by public transport (note ratios by bus, coach, hire car, taxi, train, etc.).

Space targets set down by BAA plc, similar for Heathrow Terminal 4, Gatwick North Terminal and Stansted, are as follows, stated as square metres per busy-hour passenger:

Public operational space	$20\,m^2$
Non-public operational space	$20\,m^2$
Public commercial space	$6\,m^2$
Non-public commercial space	$1.5\,m^2$
Total	$47.5\,m^2$ net

REFERENCE

Stewart, R. (2004). Class divide. *Passenger Terminal World Annual Technology Showcase Issue.*

10.3 Constraints on building form

In the 1930s multiple runways were the order of the day, but by the 1950s a pattern was emerging of single or twin runways. In other words, as growth has progressed so have the technical aids to support that growth, to the point where now a single runway can allow between 30 and 40 aircraft movements per hour, which in turn can offer an annual airport capacity of the order of 25 million passengers. A runway's capacity is determined by its independence from other neighbouring runways, the mix of aircraft and the air traffic control systems in operation. Thus, where the single runway is inadequate, the optimum of a pair of parallel, and therefore potentially independent, runways separated by at least 1600–1800 metres has developed. Such a separation allows the location of a complex of terminal buildings between the runways, with the benefit of minimal cross-runway aircraft movements. Short runway airports or STOLports (short take-off and landing), limited in use to small aircraft, are appropriate for some locations.

Obstacle clearances are laid down for both parked aircraft and buildings: a series of imaginary surfaces are defined in relation to runways and appropriate to their standards of instrumentation. These surfaces define the permissible height and position of buildings, as do lines of sight from control towers and other key installations.

Transport terminals and modal interchanges

10.4 Overall functional planning and passenger segregation in terminals

As well as the organisation of functions to suit the pattern of passenger flow (Figure 10.1), there are several possible arrangements of levels:

- *Side-by-side arrivals and departures on a single level.* Suitable for the smaller scale operations, where first-floor movement of passengers from terminal to aircraft via telescopic loading bridges is not justified.
- *Side-by-side arrivals and departures with two-level terminal.* This design obviates the need for elevated roads because all kerbside activity can take place at ground level. Escalators and lifts have to be provided to take departing passengers up to the boarding level.
- *Vertical stacking of arrivals and departures.* The majority of large-scale terminals now adopt this configuration. Departures facilities are invariably at the high level, usually accompanied by an elevated forecourt, with baggage handling and arrivals facilities below. It is essentially economic and convenient for passenger and baggage movement – departing passengers arrive at an elevated forecourt and move either on the level or down a short distance

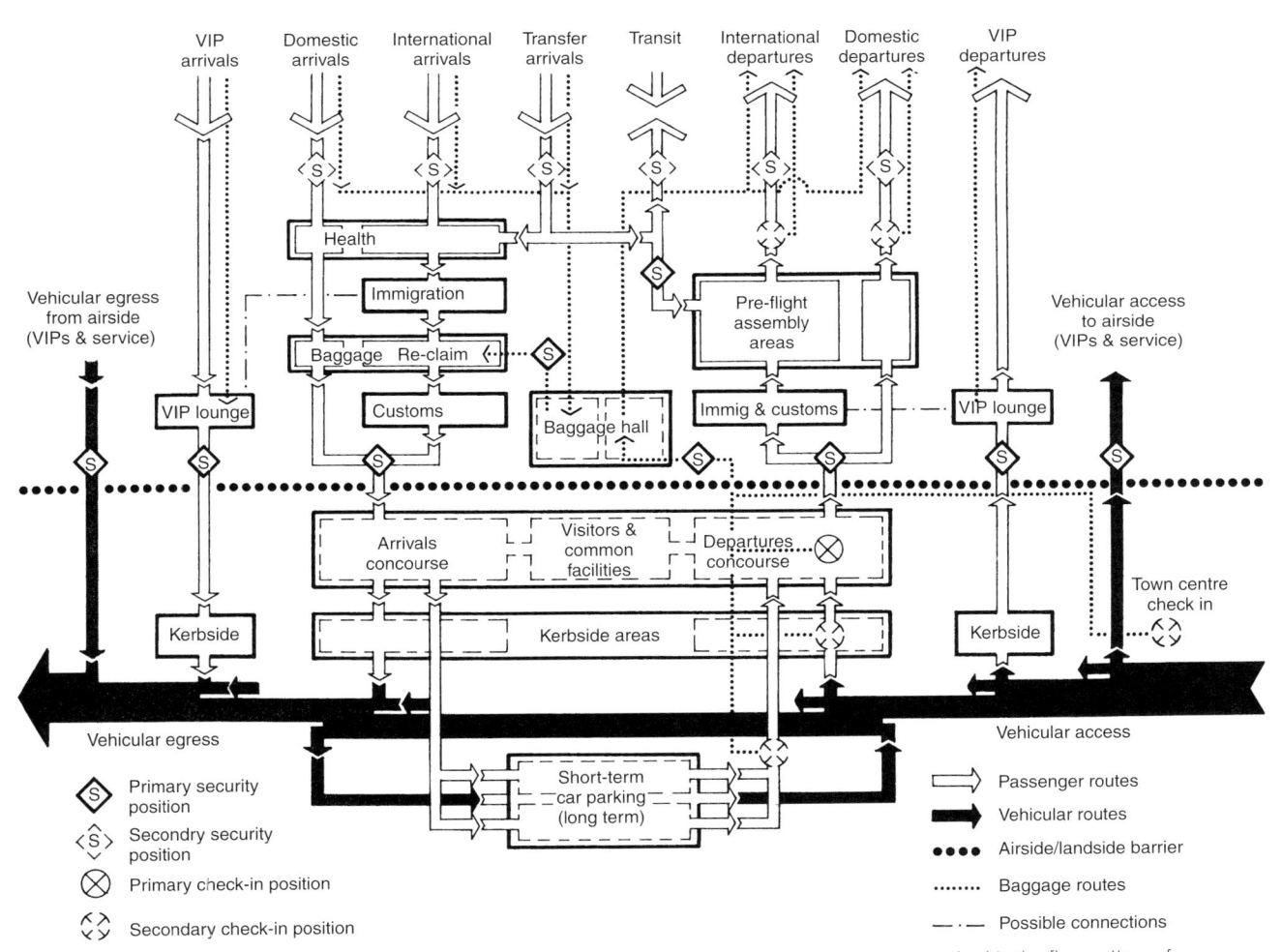

▲ 10.1 Diagram showing both passenger and vehicular flow patterns for an international plus domestic terminal.

Airport interface **175**

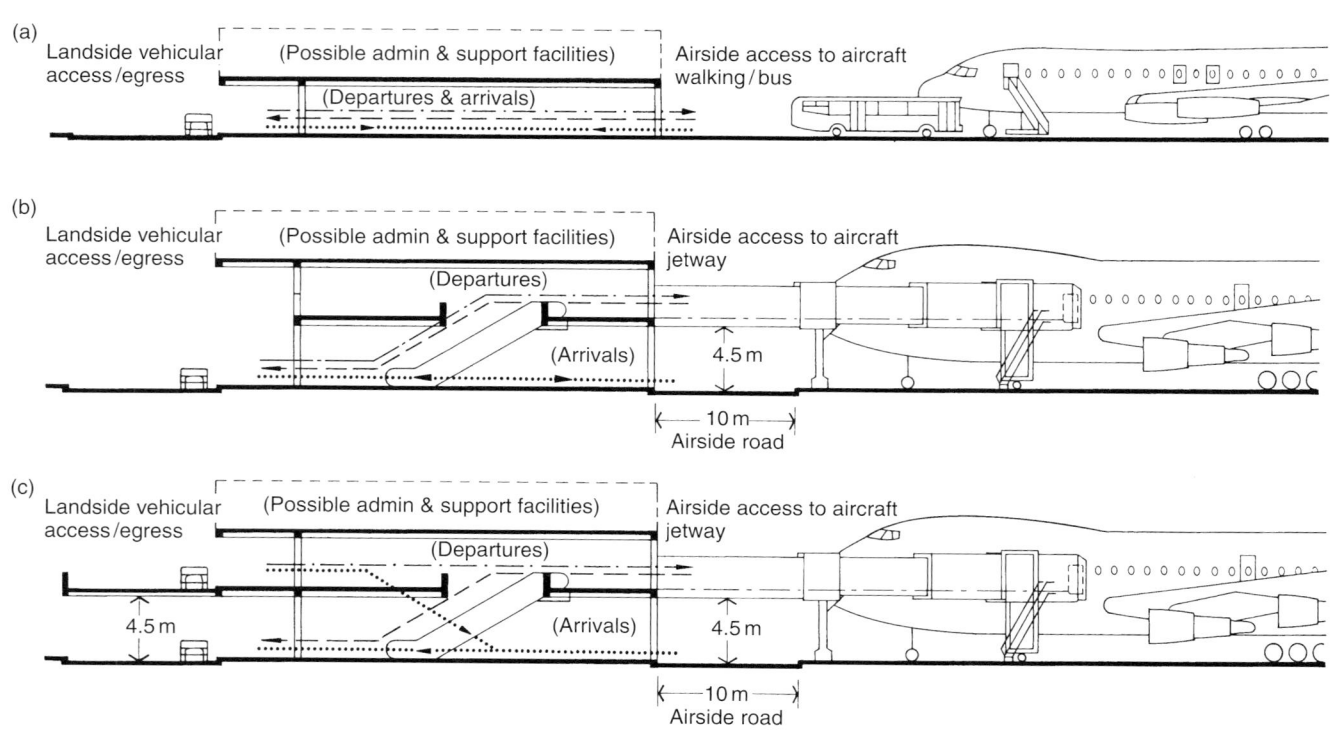

10.2 Forms of typical terminals shown by cross-sections. (a) Single-level terminal – generally applicable to small or domestic terminals. Arrival and departure routes split horizontally as flow plan diagram in Figure 10.1; (b) Two-level terminal – jetway type (horizontal split); (c) Two-level terminal – loading bridge type (vertical segregation)

by ramp to the aircraft loading point. Arriving passengers also, after leaving the aircraft, move downwards to baggage reclaim and landside facilities.

- *Vertical segregation*. High volumes of passengers, particularly with wide-bodied aircraft on long-haul routes, are best served by unidirectional circulation routes. Segregation can theoretically be either vertical or horizontal, but in practice the only feasible way to achieve it is by departing passenger routes at high level with downwards circulation to the aircraft and arriving passenger routes below.

The overall functional diagram above illustrates centralisation. Most airport terminals are centralised groups of functions, commercial, passenger and baggage processing, airline operations, etc. Centralisation has the advantage of economy of management, if not of passenger convenience. However, where absence of the need for control authorities in the case of domestic terminals or prime concern for passenger convenience at the expense of centralised control has made it possible, then decentralisation has proved beneficial.

10.5 The aircraft interface, terminal or remote parking

The number of aircraft parking places needed normally requires extended building structures in the form of piers or satellites to provide the frontage. Stands not terminal-, pier- or satellite-served have coaches to carry passengers to and from the terminal. Alternatively, superior types of bus specifically for operation on airport aprons, or even mobile lounges that raise and lower to serve terminal and aircraft door, can be used.

10.6 Landside functions

Note that rules of thumb and quantity factors quoted in the next two sections are based largely on the IATA *Airport Development Reference Manual* (2004 edition). In accordance with the recommendations therein (see Chapter 5), the level of service assumed is C.

10.6.1 Arriving at or leaving the terminal by car or public transport

Policy decisions to be applied:

- *Security.* The creation of terrorist vantage points should be avoided in the design.
- *Commercial.* For commercial reasons the whole forecourt or at least the private car section may be incorporated in the short-term or nearest car park. This will force motorists to pay for the privilege of parking close to the check-in area for more than a nominal period, which can be controlled.
- *Baggage.* For high volumes of inclusive tour traffic, with coaches setting down large pre-sorted volumes of baggage, it may be appropriate to have a dedicated area and a route to the baggage areas. Baggage trolleys should be available at intervals for passenger use.
- *Airline needs.* For large terminals shared by many airlines, it may be appropriate to have signed sections of forecourt.
- *Predicted changes.* Take account of any predictable changes in traffic mix which may affect the modal split (IATA recommendation is quoted in Chapter 5: to reduce the percentage of passengers arriving by car).

Quantities to be assessed:

- *Hourly passenger flows.* In the case of a combined departures and arrivals forecourt, a planned two-way rate will be relevant.
- *Estimated dwell time.* An average of 1.5 minutes may be allowed for cars and taxis.
- *Modal split.* Subject to local conditions, 50% of passengers may use private cars and taxis. Many types of bus and coach will call at the departures forecourt, but do not need dedicated set-down positions. In order to provide the shortest route for the greater number of passengers, coach and bus bays should be located closest to the terminal doors. However, in the case of a single-level forecourt, it may be appropriate to designate pick-up and set-down bays for specific types of bus and coach.

Typical space calculation based on 2000 originating passengers/hour:

- Number of passengers/hour at kerbside for cars + taxis: 1000.
- Number of passengers per car or taxi: 1.7 say.
- Number of cars and taxis: 1000/1.7 = 588 per hour.
- Number of cars and taxis at one time: 588/40 = 16.
- Length of kerb per vehicle: 6 m + 10%.
- Length of kerbside for cars and taxis: 105.6 m.
- Overall rule of thumb: 1.0 m of total kerbside (including public transport) per 10 passengers/hour.

10.6.2 Waiting in a landside public concourse

Policy decisions to be applied:

- *Security.* Entry to the public concourse can be controlled by a security comb, but this is the least common option, depending as it does on searching of passengers and visitors alike.
- *Commercial.* Shopping and catering facilities will be appropriate here, together with bureau de change (international terminal only), flight insurance sales office (departures), hotel bookings, car hire desks (arrivals) and post office. Provision for spectators may be made. Car park pay station for the benefit of car drivers seeing passengers off and meeting passengers.
- *Baggage.* All circulation areas should make allowance for baggage trolleys.
- *Government controls.* Access to airside to be provided for staff.
- *Airline needs.* Airlines will require ticket sales desks and offices.
- *Information systems.* Public display of information on flights. Information desk for public.
- *Predicted changes.* Provision may be made for exceptional conditions occasioned by delayed flights. Additional seating or even extra catering space, which may also be usable as airside, may be provided.

Quantity factors to be assessed:

- *Hourly passenger flow.* Two-way flow will be relevant where there is to be a combined departures and arrivals area.
- *Visitor ratio.* A common ratio in the West would be 0.5 to 0.2 visitors per passenger (with even lower ratios for certain domestic traffic) and in the East or Africa 2.5 to 6 or even higher.
- *Estimated dwell time.* A common time in arrivals would be 5 minutes for passengers and 30 minutes for meeters and greeters.

Typical space calculation based on 2000 terminating passengers/hour:

- Number of people per hour: 3400 (0.7 visitors/passenger).
- Number at one time (2000/12 + 1400/2): 867.
- Space per person (level of service C): 2.3 m^2.
- Area required: 1994 m^2.

10.6.3 Checking in, with or without baggage

Here, passengers show their tickets, have seats allocated and if necessary have large items of baggage weighed (and possibly security screened) for registration, sorting, containerisation and loading into the aircraft hold.
Policy decisions to be applied:

- *Security.* Procedures are now being introduced whereby all baggage is searched by the airline's security staff at entry to their check-in area, or by the check-in and security staff at the desk by means of X-ray units at or near the desk. The constraint is that the owner of the bag must be at hand at the moment of search in the event of a problem arising.
- *Baggage.* One or more delivery points may be required for out-of-gauge baggage.
- *Government controls.* A customs check facility for certain heavy items of baggage may be provided in the check-in area.

- *Airline needs.* Offices for airlines and handling agents will be needed, with close relationship with the check-in desks and preferably with a visual link.
- *Information systems.* Common user terminal equipment will make it possible to allocate desks to any airline at any time, thereby reducing the number of desks needed. Otherwise, the number of desks required is the sum total of those required by each handling agent.
- *Predicted changes.* The biggest single change is arising from the increase of automated ticketing and issuing of boarding passes and CUSS – common-user self-service check-in – even with baggage self-registration. Information technology which links the manual (conventional check-in system with baggage registration) and automated systems (where the passenger simply communicates with a small machine) makes it possible to reduce the number of check-in desks while retaining the necessary central control which check-in clerks have always had.

Quantities to be assessed:

- *Hourly passenger flows.* If CUTE (common-user terminal equipment) is in use, the total hourly flow to all desks can be used to compute the number. Landside transfer passengers to be included.
- *Processing rate.* A common rate would be 2.5 minutes/passenger, with faster rates for domestic passengers.
- *Estimated dwell time.* This is dependent upon the number of staffed check-in desks for a particular flight, but all check-in layouts have to make provision for queuing and a reasonable assumption is that a wait of 20 minutes is acceptable to economy class passengers.
- *Percentage of passengers using gate check-in.* This is a new facility and trends have yet to be established. Ten per cent usage of gate check-in would be a reasonable assumption where the facility is provided at all, although even there it may only be made available by the airlines and their handling agents for certain flights.

Typical space calculation based on 2000 originating passengers/hour (central check-in; this will be irrespective of the configuration of desks):

- Number of passengers per hour: 2000 excluding transfers and including gate check-in numbers.
- Number of desks: 2000/24 = 83.
- Queue depth might be 20 passengers at 0.8 m per person with check-in desks at approximately 2.0 m centres (max).
- Space per person (level of service C): 1.6 m^2 – average based on options in Chapter 7.
- Total queuing area: $83 \times 2.0 \times 16 = 2656$ m^2. Note that a discrete area is only applicable if there is a security-based separation between the landside public concourse and the check-in area.

10.6.4 Pre-departure security check

Policy decisions to be applied:

- *Baggage.* In the case of central security, baggage belonging to passengers using the gate check-in facility needs to be taken account of.

Transport terminals and modal interchanges

- *Government controls.* Security control will be the responsibility of the government/army/police force of the airport authority.
- *Airline needs.* Airlines may also wish to conduct security checks.

Quantities to be assessed:

- *Hourly passenger flows.* For central security and for gate security, allow for transfer passengers.
- *Processing rate.* X-ray units handle up to 1000 items per hour.
- *Estimated dwell time.* This is not calculable, since a problem item or passenger can very rapidly cause a queue to build up. The airport's objective must be for the security check to be carried out without interrupting the flow of passengers, but in reality staffing levels cannot totally eliminate queuing and a long queue area must be possible without interrupting access to other functions.

Typical space calculation based on 2000 originating passengers/hour (central security check):

- With two items of baggage, hand baggage and/or a coat per passenger.
- One set (one personnel metal detector + X-ray unit) handles 500 passengers per hour.
- Two thousand passengers per hour require four sets.

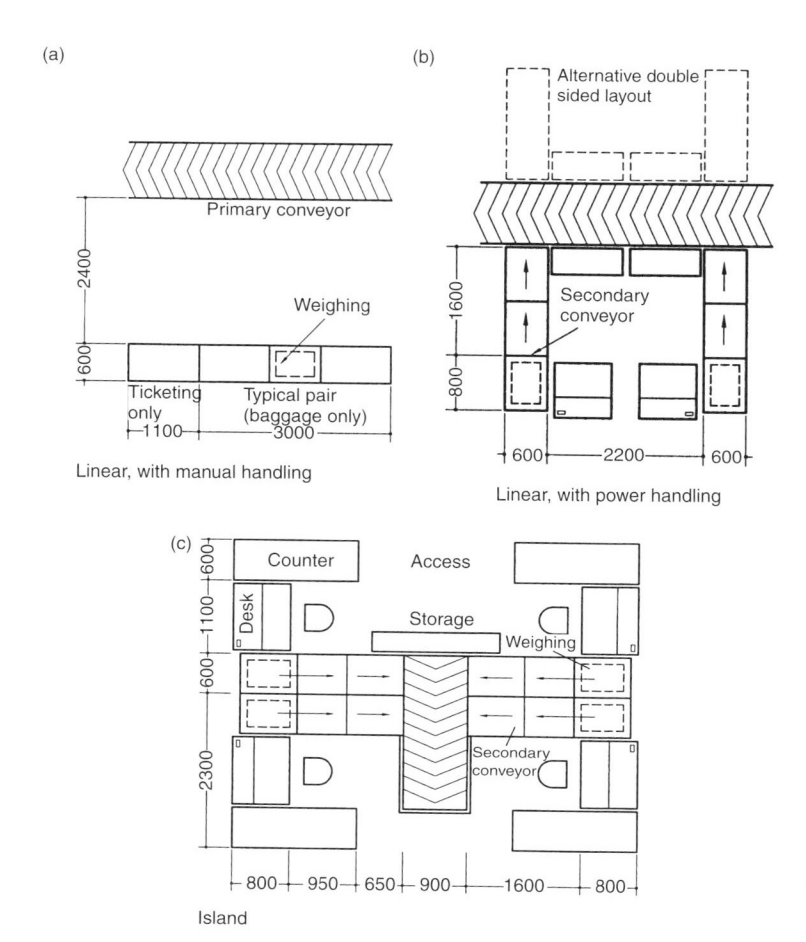

10.3 Check-in installations without security control.

Airport interface

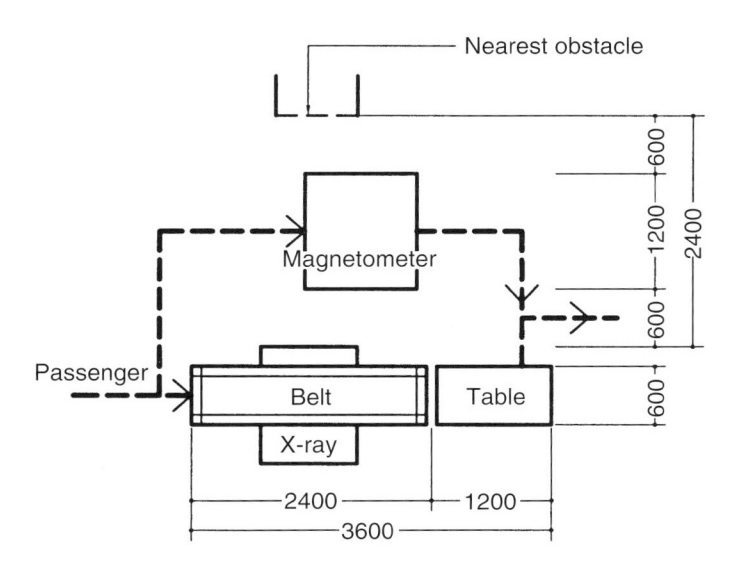

10.4 X-ray unit search of passengers and hand baggage.

10.7 Airside functions

10.7.1 Immigration check

Policy decisions to be applied:

- *Security*. A central security control brings this area under security surveillance.
- *Government controls*. Government policy will determine the designation of separate channels for different types of passport holders. Customs checks can also be carried out at this point, and offices and detention rooms will be required.
- *Predicted changes*. The changes to border controls within the European Community post-1992 are an example of the effect of international policy-making.

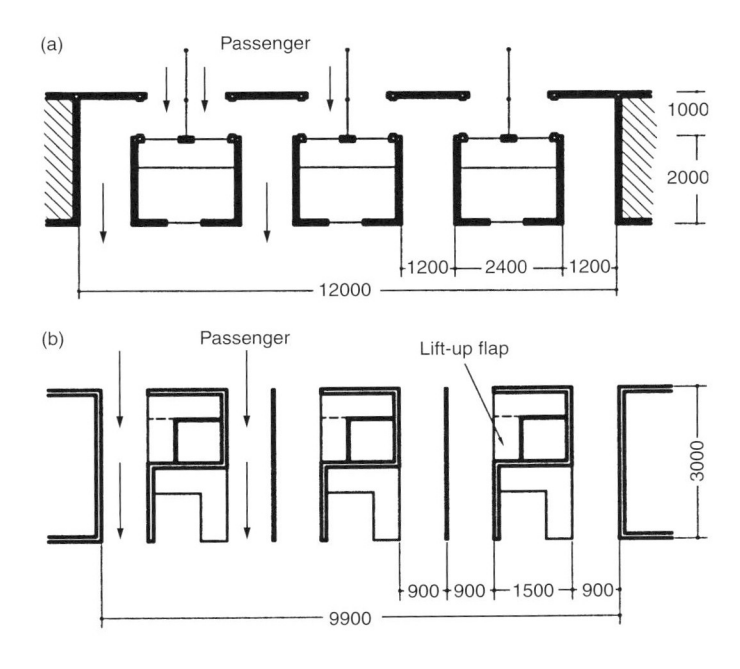

10.5 (a) Frontal presentation immigration desks, booth or open plan. (b) Side presentation immigration desks, booth or open plan (by permission of the IATA).

Airport interface

Quantities to be assessed:

- *Hourly passenger flows.* Include landside transfers.
- *Processing rate.* A common rate would be 10 seconds/passenger for departures or 30 seconds/international passenger and 6 seconds/domestic passenger for arrivals.

Typical space calculation based on 2000 originating passengers/hour:

- Number of passengers per hour: 2000 excluding transfers.
- Number of desks required: 5.5, say 6.
- Area required (25 m^2 per desk): 150 m^2.

10.7.2 Waiting in airside public concourse

Here, passengers wait, shop, eat and drink, and move sooner or later to the departure gate of their flight. In some cases, that point may be the people-mover leading to a satellite or the coach station serving remote stands.
Policy decisions to be applied:

- *Security.* If comprehensive centralised security at entry to the airside concourse is provided, no further security checks may be needed. Otherwise, security checks may be carried out at entry to an individual gate assembly area or lounge.
- *Commercial.* Shopping and catering facilities will be appropriate here, including duty-free shopping.
- *Airline needs.* Airlines will have specific requirements at the gate positions. Airlines frequently specify special lounges for first-class and business-class passengers, known as CIP (commercially important passengers) lounges.
- *Information systems.* Full information must be provided throughout the concourse, and especially at the entries, on flight numbers, departure times, delays and gate numbers.

Quantities to be assessed:

- *Hourly passenger flows.* Include landside and airside transfers.
- *Estimated dwell time.* A common standard would be 30 minutes.

Typical space calculation based on 2000 originating passengers/hour:

- Number of passengers per hour: 2000 excluding transfers.
- Number of passengers at one time: 1000.
- Space per person (level of service C): 1.5 m^2.
- Area required: 1500 m^2. For level of service C, gate holding areas should be sized to hold 65% of the maximum number of passengers boarding the largest aircraft which can dock at the stand in question.
- Space per person (level of service C): 1.5 m^2 – equivalent to a mix of seated passengers at 1.7 m^2 per person and standing at 1.2 m^2 per person.
- Area for 400-seater aircraft: 260 × 1.5 = 390 m^2.

10.7.3 Reclaiming baggage

Here, passengers wait for and reclaim their baggage, which has been unloaded from the aircraft while they have been travelling through the terminal building and passing through the immigration control.
Policy decisions to be applied:

- *Baggage.* There needs to be a means of delivering out-of-gauge baggage to the passengers, and also a means of passengers claiming their

baggage after they have passed through to the landside, either because they have forgotten it or because, due to airline problems, it has arrived on a different flight from them.
- *Information systems*. Numbers of reclaim units need to be displayed against the arriving flight numbers, particularly in areas where passengers are entering the reclaim area.

Quantities to be assessed:

- *Hourly passenger flows*. Landside transfer passengers need to reclaim their baggage.
- *Processing rate*. There are several ways of calculating throughput in baggage reclaim. A reclaim device for narrow-bodied aircraft should have a length of 30–40 m and one for a wide-bodied aircraft should have a length of 50–65 m. Average occupancy times for narrow- and wide-bodied aircraft would be 20 and 45 minutes respectively.
- *Estimated dwell time*. A common standard would be 30 minutes.
- *Number of checked-in bags per passenger*. Possibly an average of 1.0 depending on whether the flight is long haul or short haul, although the flow calculation method used does not depend upon this factor. (See Figure 10.6.)

Typical space calculation based on 2000 terminating passengers/hour:

- Number of passengers per hour: 2000 excluding transfers.
- Number of passengers at one time: 1000.
- Space per person (level of service C): 2.3 m^2.
- Area required: 2300 m^2. However, the operative calculation is for the number of reclaim units and the space round each for a flight load of passengers waiting – assume 50% of passengers arrive by wide-bodied and 50% by narrow-bodied aircraft.
- Number of passengers per narrow-bodied aircraft at 80% load factor: 100.
- Number of passengers per wide-bodied aircraft at 80% load factor: 320.
- Number of narrow-bodied devices: 1000/3 × 100 = 4.
- Number of wide-bodied devices: 1000/1.33 × 320 = 3.
- Space per person in retrieval and peripheral area around unit (level of service C): 1.7 m^2.
- Retrieval and peripheral area around 40-metre narrow-bodied device: 177 m^2 (100 persons).
- Retrieval and peripheral area around 65-metre wide-bodied device: 270 m^2 (160 persons).
- Total waiting area: (4 × 177 + 3 × 270) plus 50% aisles = 2265 m^2 (excluding lateral circulation area within baggage reclaim area and trolley storage areas).

10.7.4 Inbound customs clearance

Policy decisions to be applied:

- *Security*. Customs officers are increasingly on the lookout for narcotics rather than contraband.
- *Government controls*. Offices and search rooms will be required. Type of surveillance will need to be determined.

Transport terminals and modal interchanges

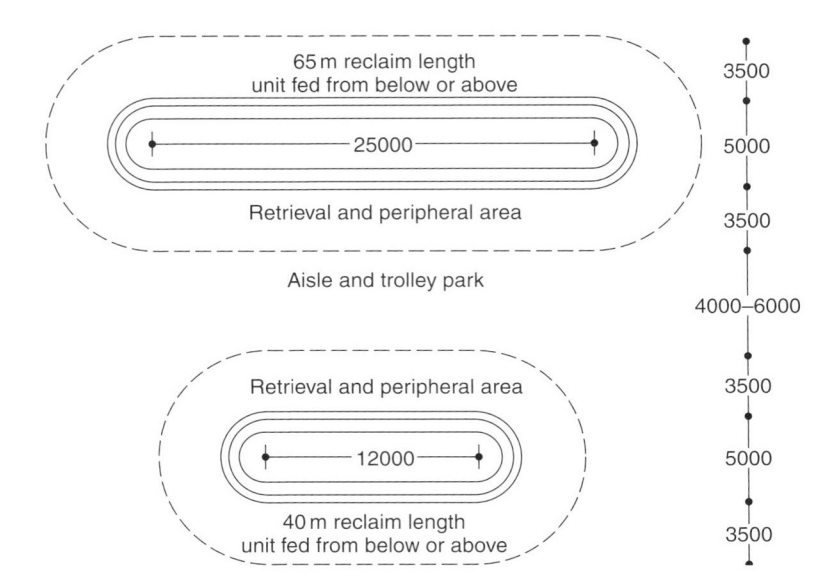

```
                  65 m reclaim length
               unit fed from below or above          3500
                ┌──────────────────────┐
                │   ●────25000────●    │             5000
                └──────────────────────┘
                Retrieval and peripheral area        3500

                    Aisle and trolley park
                                                     4000–6000

                Retrieval and peripheral area        3500
                ┌──────────────────────┐
                │   ●────12000────●    │             5000
                └──────────────────────┘
                   40 m reclaim length
                unit fed from below or above         3500
```

10.6 Typical baggage reclaim configuration.

- *Predicted changes*. The changes to border controls within the European Community post-1992 and the introduction of the blue channel for EC passengers moving freely between member states are an example of the effect of international policy-making.

Quantities to be assessed:

- *Hourly passenger flows*. Include landside transfers.
- *Processing rate*. A rate for passengers being searched would be 2 minutes/passenger.

Typical space calculation based on 2000 terminating passengers/hour:

- Area required if rule of thumb is 0.5 m^2 per passenger per hour: 1000 m^2.

10.8 Aircraft and apron requirements

10.8.1 Baggage handling

The manoeuvring of trains of baggage containers and trailers determines the layout of baggage loading and unloading areas. The sorting and security screening of baggage is not a subject covered by this book.

10.8.2 Loading bridges

A range of types of loading bridges is available. These are otherwise known as air-bridges, air-jetties or jetways, which connect terminal to aircraft.

A strong determinant of airside space as well as passenger numbers in the terminal buildings themselves is the design and size of aircraft. 2007 will see for some large airports the advent of the biggest change in aircraft size since the roll-out of the first Boeing 747 in September 1968: the Airbus 380. Most significantly, this double-decker aircraft has a wingspan of nearly 80 metres and a tail height of 24 metres. ICAO has created a new Code F for airport dimensions, and plans are under way for loading equipment and terminal modifications.

Airport interface

Transport terminals and modal interchanges

▶ 10.7 Baggage handling transport: double container dolly.

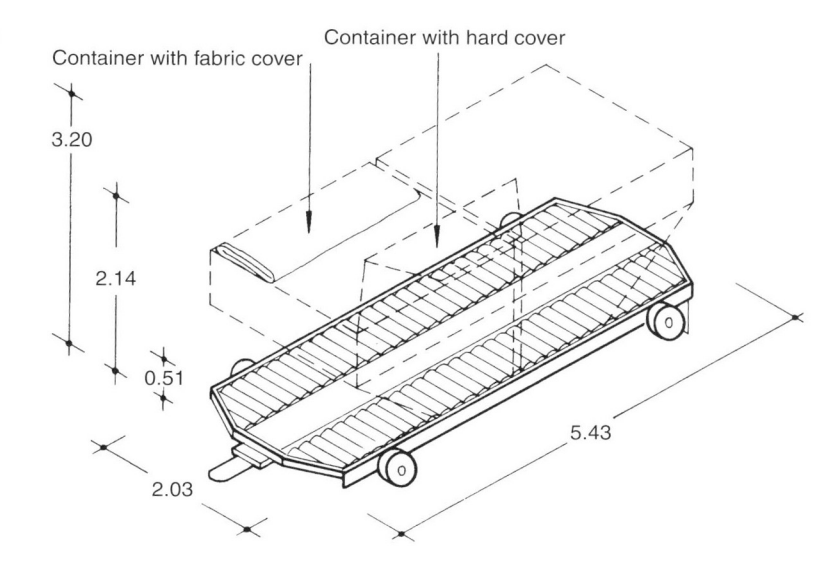

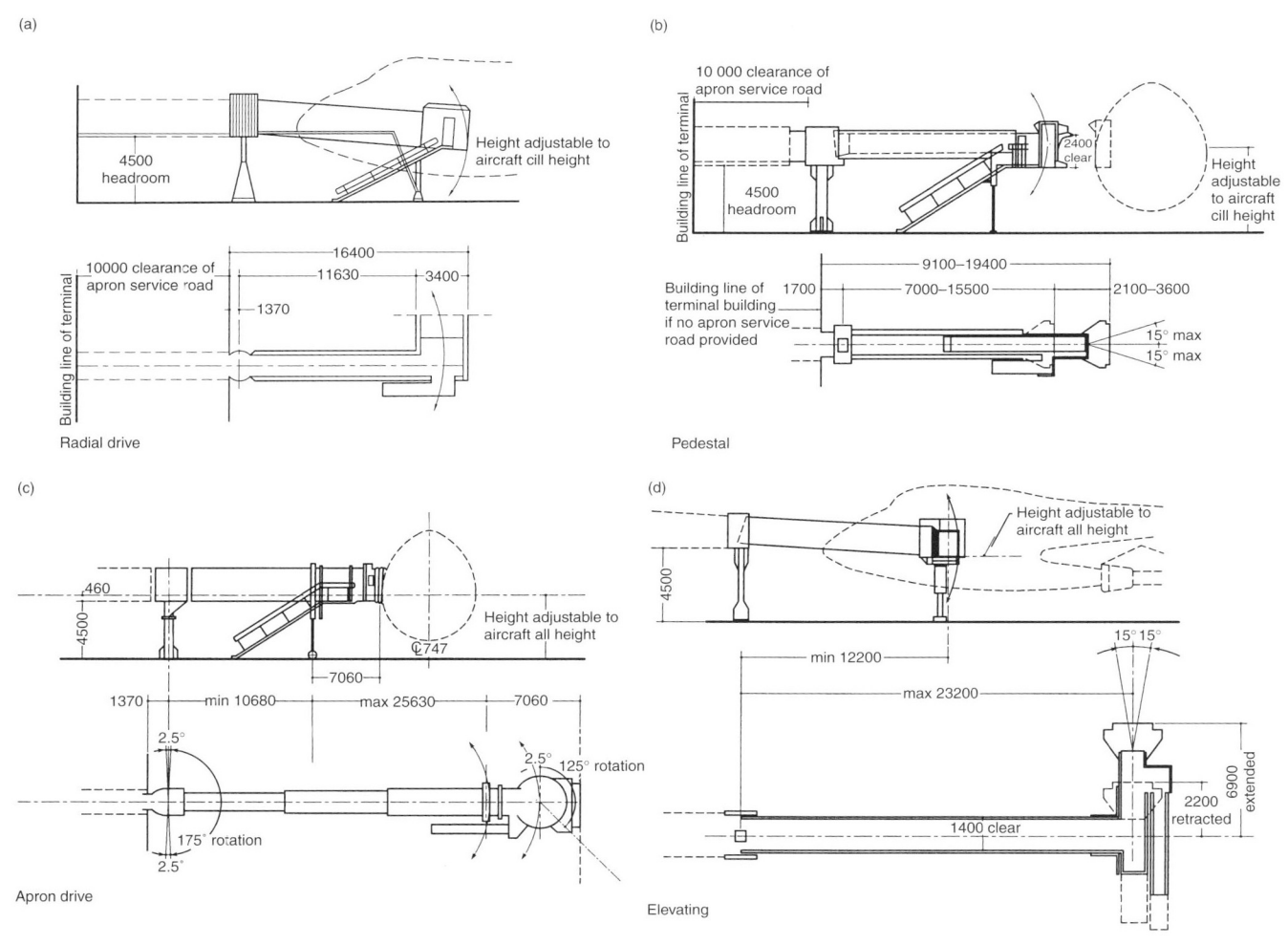

▲ 10.8 Four loading bridge types: plans and elevations.

(a) Radial drive

(b) Pedestal

(c) Apron drive

(d) Elevating

Airport interface **185**

Table 10.1 ICAO codes (from Annex 14 of the Convention on International Civil Aviation, 2004; stand sizes, Codes E, D, C and B)

Code	Aircraft	Stand depth* (m)	Stand width* (m)
E	B747 range, B777 and A330/A340	70.66	65.00
D	Suitable for all sizes between MD11 and A310 inclusive (i.e. DC10/MD11, A300, B767, L1011, B757, A310)	61.21	52.00
C	B727, MD80/90, A320, B737, BAC1-11, BAe146, F28/100/27/50, ATR42/72, ATP, Dash 7 and 8	46.69	36.00
B	Suitable for smaller turboprops only	22.00	24.00

*Excluding positional tolerance.

REFERENCES

International Civil Aviation Organisation (2004). Annex 14 to the Convention on International Civil Aviation. ICAO.

International Civil Aviation Organisation (2004). *Airport Planning Manual*. ICAO.

Table 10.2 Aircraft dimensions (this listing is based on published data on the principal civil passenger airliner types current in 2003)

Narrow-bodied jet transport aircraft
The following 15 aircraft types account for over 13 000 civilian aircraft currently using the world's airports, ranging in size from the BAe 146 to the Boeing 757. Former USSR types omitted.

Airliner: Airbus A320–200
Nationality: European
Number manufactured: 1220 orders by 2000 for 320, and 603 for shortened A319 and 264 for stretched A321
Number of passengers: up to 179 (6 abreast)
Wingspan: 33.91 m (111 ft 3 in)
Length: 37.57 m (123 ft 3 in), A319 is 33.80 m (110 ft 11 in), A321 is 44.51 m (146 ft 0 in)
Height: 11.80 m (38 ft 9 in)

Airliner: Boeing 727–100 and 200
Nationality: USA
Number manufactured: 1831 (last delivery 1984)
Number of passengers: up to 189 (6 abreast)
Wingspan: 32.92 m (108 ft 0 in)
Length: 46.69 m (153 ft 2 in). Note that 727–100 is 6 m shorter
Height: 10.36 m (34 ft 0 in)

Airliner: Boeing 717–200 (ex MD-95)
Nationality: USA
Number manufactured: in service Sept 1999
Number of passengers: up to 106
Wingspan: 28.47 m (93 ft 5 in)
Length: 37.80 m (124 ft 0 in)
Height: 8.86 m (29 ft 1 in)

Airliner: Boeing 737–100 and –200
Nationality: USA
Number manufactured: 1144 (last delivery 1987)
Number of passengers: up to 130 (6 abreast)
Wingspan: 28.35 m (93 ft 0 in)
Length: 29.54 m (96 ft 11 in) Note that 737–100 is shorter
Height: 11.28 m (37 ft 0 in)

Airliner: Boeing 737–300
Nationality: USA
Number manufactured: 1108 orders by June 1999
Number of passengers: up to 149 (6 abreast)
Wingspan: 28.88 m (94 ft 4 in)
Length: 33.40 m (109 ft 7 in)
Height: 11.13 m (36 ft 6 in)

Table 10.2 (Continued)

Airliner: Boeing 737–400 (a stretched 737–300)
Nationality: USA
Number manufactured: 486 orders by June 1999
Number of passengers: up to 156 (6 abreast)
Wingspan: 28.88 m (94 ft 4 in)
Length: 36.45 m (119 ft 7 in)
Height: 11.13 m (36 ft 6 in)

Airliner: Boeing 737–500 (a short-body 737–300)
Nationality: USA
Number manufactured: 388 orders by June 1999
Number of passengers: up to 132 (6 abreast)
Wingspan: 28.88 m (94 ft 4 in)
Length: 31.01 m (101 ft 9 in)
Height: 11.13 m (36 ft 6 in)

Airliner: Boeing 737–600, –700 and –800 (new generation)
Nationality: USA
Number manufactured: 1100 orders by June 1999
Number of passengers: 108 (–600), 128 (–700) and 160 (–800) (6 abreast)
Wingspan: 34.31 m (112 ft 7 in)
Length: 33.63 m (110 ft 4 in) of –700, –600 being shorter and –800 being 39.48 m
Height: 12.55 m (41 ft 2 in)

Airliner: Boeing 757–200 and –300 (new 1999 stretched version)
Nationality: USA
Number manufactured: 872 orders by August 1999
Number of passengers: up to 239 (6 abreast), or up to 289 (–300)
Wingspan: 38.05 m (124 ft 10 in)
Length: 47.32 m (155 ft 3 in), –300 is stretched by 7.13 m
Height: 13.56 m (44 ft 6 in)

Airliner: British Aerospace BAe 146–100, –200 and –300, now renamed Avro RJ series.
Nationality: UK
Number manufactured: 219 BAe 146 sales up to 1993. 154 RJ orders up to September 1999.
Number of passengers: up to 112 (6 abreast)
Wingspan: 26.34 m (86 ft 5 in)
Length: 30.99 m (101 ft 8 in) of –300, –100 and –200 being shorter
Height: 8.61 m (28 ft 3 in)

Airliner: Canadair Regional Jet CRJ100/CRJ200/CRJ700
Nationality: Canada
Number manufactured: 313 deliveries by June 1999
Number of passengers: up to 50 (4 abreast)
Wingspan: 21.21 m (69 ft 7 in)
Length: 26.77 m (87 ft 10 in), 70-seater CRJ700 is 32.87 m (107 ft 10 in)
Height: 6.22 m (20 ft 5 in)

Airliner: Embraer ERJ–135, –145, –170/190
Nationality: Brazil
Number manufactured:
Number of passengers: up to 100
Wingspan: 28.08 m (92 ft 1 in)
Length: 36.24 m (118 ft 11 in) max is for 170/190

Airliner: Fokker F28 and F70/100
Nationality: Dutch
Number manufactured: 241 F28 (pre-1986), and 325 F70/100
Number of passengers: up to 119 (5 abreast)
Wingspan: 28.08 m (92 ft 1 in) F28 is smaller span and length
Length: 35.31 m (115 ft 10 in), 79-seater F70 is 30.91 m long
Height: 8.60 m (27 ft 11 in)

Airliner: McDonnell Douglas MD-80 series (successor to DC-9)
Nationality: USA, also China (assembly agreement 1985)
Number manufactured: 1192 deliveries by 2000
Number of passengers: up to 172 (5 abreast)
Wingspan: 32.87 m (107 ft 10 in)
Length: 45.06 m (147 ft 10 in) MD-87 is short version, length 39.70 m (130 ft 5 in)
Height: 9.04 m (29 ft 8 in)

Table 10.2 (Continued)

Airliner: McDonnell Douglas MD-90 (successor to MD-80)
Nationality: USA, also China (assembly agreement 1992)
Number manufactured: 114 by 2000
Number of passengers: up to 172 (5 abreast)
Wingspan: 32.87 m (107 ft 10 in)
Length: 46.51 m (152 ft 7 in)
Height: 9.33 m (30 ft 7 in)

Wide-bodied jet transport aircraft
The following 15 aircraft types, which each carry over 280 passengers, accounted in 1999 for 4000 civilian aircraft using the world's airports. They range in size up to the present Boeing 747–400, with a wingspan of over 64 metres and a fuselage of over 70 metres, with the A380 to come, with a wingspan of nearly 80 metres and a fuselage of over 72 metres. Former USSR types omitted.

Airliner: Airbus A300 various versions
Nationality: European
Number manufactured: 520 orders up to Sept 1999
Number of passengers: up to 344 (up to 9 abreast)
Wingspan: 44.84 m (147 ft 1 in)
Length: 54.08 m (177 ft 5 in)
Height: 16.62 m (54 ft 6 in)
Size data for A300–600: note that 300B2 and 300B4 are shorter

Airliner: Airbus A310
Nationality: European
Number manufactured: 261 orders by Sept 1999
Number of passengers: up to 280 (up to 9 abreast)
Wingspan: 43.90 m (144 ft 0 in)
Length: 46.66 m (153 ft 1 in)
Height: 15.81 m (51 ft 10 in)

Airliner: Airbus A330–300 (2 engines) and A340–300 (4-engine version)
Nationality: European
Number manufactured: 530 orders for A-330 and A-340 by Sept 1999
Number of passengers: up to 440 (up to 9 abreast)
Wingspan: 60.03 m (197 ft 10 in)
Length: 63.65 m (208 ft 10 in)
Height: 16.74 m (54 ft 11 in)

Airliner: Airbus A330- and 340–200 (longer range version of A330- and 340–300)
Nationality: European
Number manufactured:
Number of passengers: up to 303 (up to 9 abreast)
Wingspan: 60.03 m (197 ft 10 in)
Length: 59.39 m (194 ft 10 in)
Height: 16.74 m (54 ft 11 in)

Airliner: Airbus A380–800
Nationality: European
Number manufactured: 121 orders by Sept 2003, deliveries in 2007
Number of passengers: Airbus suggest 555 (optimum 10 abreast)
Wingspan: 79.80 m (261 ft 10 in)
Length: 72.70 m (238 ft 6 in)
Height: 24.10 m (79 ft 1 in)

Airliner: Boeing 747–100, 200 and 300
Nationality: USA
Number manufactured: 724
Number of passengers: up to 516 (Srs-300 624) (10 abreast)
Wingspan: 59.64 m (195 ft 8 in)
Length: 70.67 m (231 ft 10 in)
Height: 19.30 m (63 ft 4 in)

Airliner: Boeing 747–400
Nationality: USA
Number manufactured: 464 orders by 1999
Number of passengers: up to 660 (11 abreast)
Wingspan: 64.67 m (212 ft 2 in)

Table 10.2 (Continued)

Length: 70.67 m (231 ft 10 in)
Height: 19.30 m (63 ft 4 in)

Airliner: Boeing 747SP
Nationality: USA
Number manufactured: 43
Number of passengers: up to 440 (11 abreast)
Wingspan: 59.64 m (195 ft 8 in)
Length: 56.31 m (184 ft 9 in)
Height: 19.94 m (65 ft 5 in)

Airliner: Boeing 767–200
Nationality: USA
Number manufactured: 239 orders by 1999
Number of passengers: up to 290 (8 abreast)
Wingspan: 47.57 m (156 ft 1 in)
Length: 48.51 m (159 ft 2 in)
Height: 15.85 m (52 ft 0 in)

Airliner: Boeing 767–300
Nationality: USA
Number manufactured: 540 orders by 1999
Number of passengers: up to 330 (8 abreast)
Wingspan: 47.57 m (156 ft 1 in)
Length: 54.94 m (180 ft 3 in)
Height: 15.85 m (52 ft 0 in)

Airliner: Boeing 777–200
Nationality: USA
Number manufactured: 429 by 1999
Number of passengers: up to 440 (10 abreast)
Wingspan: 60.95 m (199 ft 11 in) note folding wing option reduces
Length: 63.73 m (209 ft 1 in)
Height: 18.45 m (60 ft 6 in)

Airliner: Lockheed L-1011–100 and 200 Tristar
Nationality: USA
Number manufactured: 249 (including 500 series) (last delivery 1984)
Number of passengers: up to 400 (10 abreast)
Wingspan: 47.34 m (155 ft 4 in)
Length: 54.17 m (177 ft 8 in)
Height: 16.87 m (55 ft 4 in)

Airliner: Lockheed L-1011–500 Tristar
Nationality: USA
Number manufactured: see 100 and 200 series data
Number of passengers: up to 330 (10 abreast)
Wingspan: 50.09 m (164 ft 4 in)
Length: 50.05 m (164 ft 2 in)
Height: 16.87 m (55 ft 4 in)

Airliner: McDonnell Douglas DC-10 series 30 (also series 10)
Nationality: USA
Number manufactured: 386 (last delivery 1989)
Number of passengers: up to 380 (10 abreast)
Wingspan: 50.40 m (165 ft 4 in); series 10 is less
Length: 55.50 m (182 ft 1 in); series 10 is 0.35 m longer
Height: 17.70 m (58 ft 1 in)

Airliner: McDonnell Douglas MD-11 (successor to DC-10)
Nationality: USA
Number manufactured: 136 (last delivery 2000)
Number of passengers: up to 405 (10 abreast)
Wingspan: 51.70 m (169 ft 6 in)
Length: 61.21 m (200 ft 10 in)
Height: 17.60 m (57 ft 9 in)

Turboprop transport aircraft
Not included in this table

11 Twenty-first-century trends

Initiatives are one thing. All the initiatives mentioned in Chapter 5, and many others, are redefining the rules.

Visible trends are another. Airports are getting their rail links and rising in some cases to 'city' status.

11.1 Evolution of transport in relation to the city

Evolution in relation to transport for cities and, to a lesser extent, towns:

Pre-industrial	zoning, Ebenezer Howard, the family orientation
19th century rail	'crashing' into city
City remote 20th century	airport incompatible, therefore remote
City connected late 20th century	city–airport corridor of transport
City integrated 21st century	airport as part of the city

(from a paper at Passenger Terminal Expo 2002 by David Holm of Woodhead International)

The above summary can be reapplied specifically to rail and bus transport. Evolution in relation to rail and bus transport for cities and, to a lesser extent, towns:

Pre-industrial	zoning, Ebenezer Howard, the family orientation
19th century rail	'crashing' into city
City first half 20th century	railways stable
City dis-integrated late 20th century	incompatible airport remote, railways in decline
City connected late 20th century	corridor of transport
City integrated 21st century	airport and railway options as part of the city

REFERENCE

Holm, D. (2002). Presentation at Passenger Terminal Expo, Cannes.

11.2 Problems affecting transportation, 2000–2025

All the following will in turn affect the future of the interchange:

- Projected growth. Irrespective of absolute growth, the aim of many airports is to ensure that 50% of passengers use public transport and to reduce the growth in car parking at airports.

- Demand on infrastructure caused by longevity and overdue maintenance.
- Demand on infrastructure arising purely from growth projections.
- Imposition of limits on pollutants and noise.
- Demands for security.
- Opposition from 'green' lobby and planning procedures.
- Cost and problems of sharing investment between public and private sectors.
- Finite limits to oil supplies, cost of fuel and volatility of supply.
- Investment in aircraft and railway rolling stock conditioned by commercial factors as well as the preceding eight.

11.3 Converging standards at the interchange

11.3.1 New expectations

Once air fares are competing with rail fares or even coach fares, then facilities can or should be comparable. Whereas an airline passenger before the days of no-frills or low-cost carriers would spend 15–20 Euros in the terminal, the low-cost flier might be spending a tenth of that.

11.3.2 The train that thinks it's a plane

Rail–air substitution is a reality in Europe, either for airport-to-airport journeys or for city-centre-to-city-centre journeys. High-speed trains in Europe are linking airports directly to cities which would otherwise have been connected by air journeys and train connections to city centre.

Forty per cent of journeys in Europe are less than 500 km, so can be made by train, reducing pressure on airports. Note that the advent of high-speed rail from London to the Channel Tunnel, and thence to Paris, Amsterdam and Brussels, is predicted to take significant traffic from airlines (see Chapter 6.6.5).

11.3.3 Don't like buses? It's a tram by any other name

> A new type of bus that looks more like a tram is being designed by First Group in a drive to get more passengers out of their cars and on to public transport. The bus and rail operator said yesterday that research showed people preferred trams to buses, so it had decided to build a 'tram bus'.
>
> First Group plans to introduce the vehicle next year as part of its 'Yorkshire showcase' of new bus services for major conurbations, including Leeds, Bradford and Sheffield.
>
> David Leeder, First Group's UK bus chief, said: 'We are looking at replacing conventional bus services with a more tram-like vehicle running on rubber tyres rather than rails'. Moir Lockhead, First Group chief executive, said the new vehicle would cost 'substantially less' than trams, which cost '£2m per vehicle before you start laying the track'.
>
> The low-floor vehicle would also be more flexible than trams, which are confined to set routes. The National Audit Office found last month that passenger forecasts on tram routes often proved optimistic.
>
> (Osborne, 2004)

11.3.4 Information technology

No longer is technology merely supplying information and regulating activity inside the airport terminal, where display screens, screening devices and information technologies have been operating for many years. The same services are seen on the railway station platform and even the urban bus-stop. It is in the screening that great change is now coming with the advent of biometrics.

REFERENCES

Calder, S. (2003). *No Frills*. Virgin Books.

Osborne, A. (2004). Don't like buses? It's a tram by any other name. *Daily Telegraph*, 13 May, p. 33.

11.4 Commercial motives for development of interchanges

- *Private investment.* While railways are almost everywhere seen as a publicly-owned national asset, private development of railway lands and stations as well as airports has romped ahead. Most airports now have or are planning private participation.
- *Filling in the gaps.* For example, 304 London Underground stations, few of which are to current standards except the new Jubilee Line stations. Many await private sector initiatives. Similarly, many towns and cities have stations awaiting development, such as Guildford (Figures 11.1 and 11.2) and Farnborough (Figure 11.3).

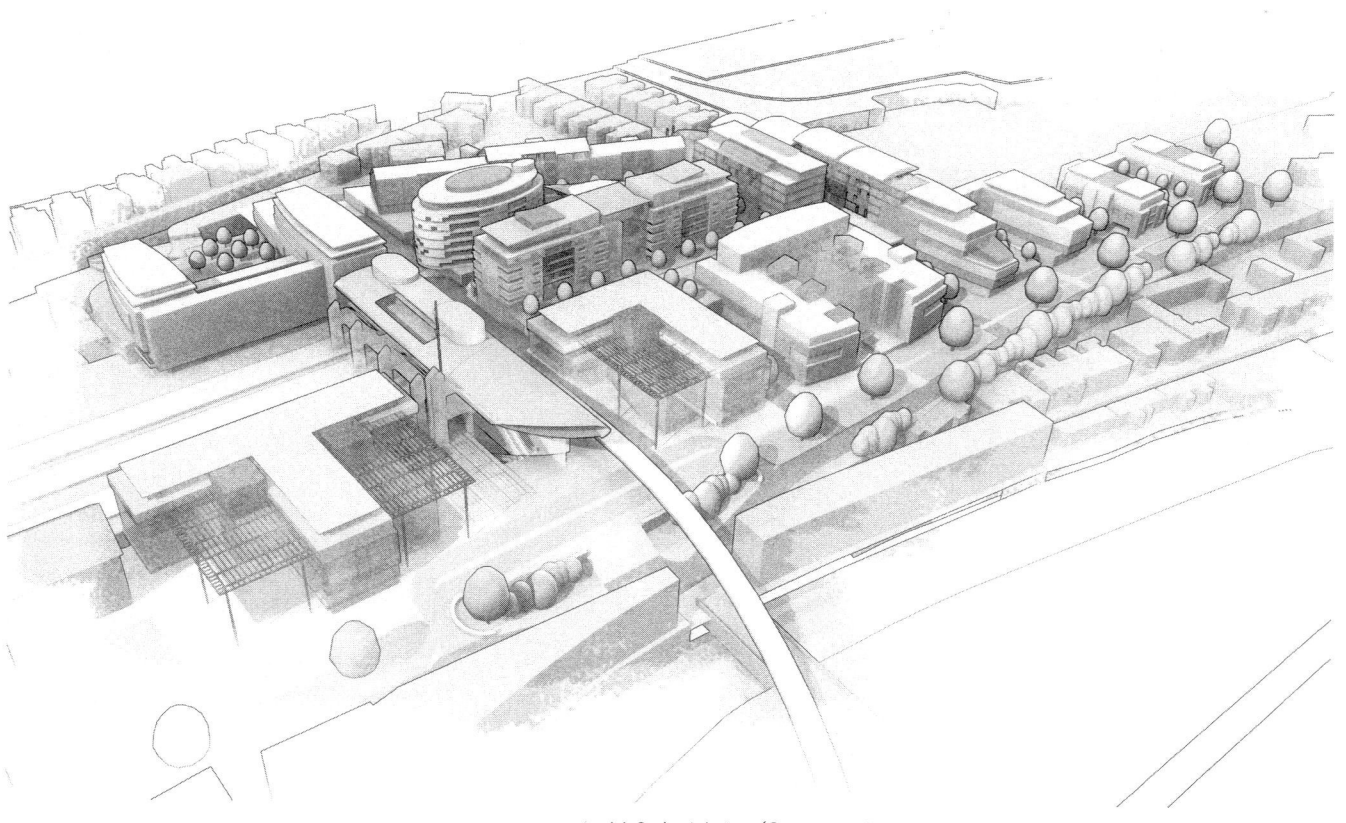

▲ 11.1 Guildford Station (Courtesy of client, Crest Nicholson Properties, and architects, Scott Brownrigg; artist D. Penney).

▲ 11.2 Aerial view (Courtesy of client, Crest Nicholson Properties, and architects, Scott Brownrigg).

▶ 11.3 Farnborough Station (courtesy of client, Network Rail, and architects, Scott Brownrigg; artist B. Minney).

REFERENCE

Ashford, N. and Moore, C.A. (1999). *Airport Finance*. The Loughborough Airport Consultancy.

11.5 Reclamation of the interchange: social, commercial and sustainable

- Shabby, unconnected bus and railway stations should be a thing of the past. Bus and railway stations will emulate standards at airports.
- Disconnection at the interchange will be remedied, and an urban connector will join the different modes of public transport – air, rail and road.
- Inaccessibility of the interchange will be remedied and urban ribbons will radiate from the transport node.
- Unattractive public spaces at the interchange will be remedied and the transport node will be a community hub.
- Low-density unproductive land at the interchange will be remedied and enhanced commercial and social potential will result.

Much more attractive public transport and interchanges will limit the need for unsustainable personal transport.

Bibliography

Adler, D. (ed.) (1999). *Metric Handbook*, 2nd edition. Architectural Press.

Ashford, N. and Moore, C.A. (1999). *Airport Finance*, 2nd edition. Loughborough Airport Consultancy.

Binney, M. (1999). *Airport Builders*. Academy Editions.

Blow, C.J. (1996). *Airport Terminals*, 2nd edition. Architectural Press.

Bode, S. and Millar, J. (1997). *Airport.* The Photographers' Gallery. Essays: *Lost in Space* by John Thakara, *Hubs* by Douglas Coupland, and *Airport Town* by P. Andreu (interview).

Buang, A. (ed.) (1998). *The Making of KLIA*. KL International Airport Berhad, Kuala Lumpur. The story of Sepang Airport.

Cuadra, M. (2002). *World Airports*. Junius Verlag GmbH, Hamburg.

Dempsey, P.S. (1999). *Airport Planning & Development Handbook*. McGraw-Hill.

De Neufville, R. and Odoni, A.R. (2003). *Airport Systems, Planning, Design and Management.* McGraw-Hill.

Doyle, J. (ed.) (1998). *From Concept to Take-off.* JOEM PR (Far East), Hong Kong. The story of Chek Lap Kok Airport in pictures.

Edwards, B. (1997). *The Modern Station*. Spon.

Edwards, B. (1998). *Modern Terminals*. Spon.

Fentress Bradburn Architects (2000). *Gateway to the West.* Images Publishing Group, Melbourne, Australia. The story of Denver Airport.

Fielden, G.B.R. Wickens, L.A.H. and Yates, I.R. (eds) (1995). *Passenger Transport after 2000AD*. Spon.

International Air Transport Association (2004). *Airport Development Reference Manual*, 9th edition. IATA.

JSK (Joos, Schulze, Krüger-Heyden) (2001). *Airports*. Ernst Wasmuth, Tübingen.

Parissien, S. (1997). *Station to Station*. Phaedon.

Powell, K. (2001). *The Jubilee Line Extension*. Lawrence King Publishing.

Quartermaine, P. (1999). *Port Architecture*. Academy Editions.

Richards, B. (2001). *Transport in Cities*. Spon.

Ross, J. (2000). *Railway Stations, Planning, Design and Management.* Architectural Press.

Sheppard, C. (1996). *Railway Stations: Masterpieces of Architecture*. Todtri Productions, New York.

Thorne, M. (ed.) (2001). *Modern Trains and Splendid Stations*, exhibition catalogue 2001–2002. Messell Publishers/The Art Institute of Chicago.

Zukowsky, J. (ed.) (1996). *Buildings for Air Travel.* Art Institute of Chicago/Prestel.

No author (1999). *Transport Spaces*, Volume 1. Images Publishing Group, Melbourne, Australia.

Index

Note: Page numbers of illustrations are given in italics

Access, 8
Aedas Architects, 96
Aer Rianta, 120
Aeroports de Paris, 62–4
AiRail, Frankfurt, 65–9, *67–9*
Aircraft:
   and apron requirements, 184–9
   and stand dimensions, 184–9
   interface, 176
Airport:
   airside functions, 181–4
   check-in, 178, *180*
   interface, 173–89
   landside functions, 177–80
   passenger and vehicle flow, *176*
   runways, 174
   terminal capacity and size, 173
   terminal forms, *176*
   terminal planning, 173
Airport/railway interchange:
   contiguous, 62–81
   linked adjacent, 82–115
   remote, 116–22
   vertical separation, 38–61
Airports Council International, 164
   worldwide traffic figures, 38
Airtrack, 48
Alsop Architects, 149
Amsterdam Schiphol Airport, 43–5, *44–5*
Arup:
   Heathrow Terminal 5, 50
   Hong Kong, 76
   Sydney, 158
   Zurich, 38
Ashford International Station, Kent, 127–30, *128–30*
Atlanta Hartsfield–Jackson Airport, 108, *108*

BAA (the former British Airports Authority):
   Heathrow Terminal 4, 51
   Heathrow Terminal 5, 50
   Gatwick, 74
   Southampton, 106
   Space targets, 174
   Stansted, 73
   Surface Access Strategy, 35
Baggage handling, 7, 178, 182, *184*, 184, *185*
BART, San Francisco, 111
Benthem Crouwel Architekten, 45
Birmingham Airport, 86–91, *88–91*
Birmingham International Airport Ltd, 86–7
Boeing aircraft, 7
Bombardier, *112*

Border controls, 163
Brookes Stacey Randall, 148
Building design legislation, 163
Building Design Partnership, 134
Bus and coach interface, 165–70
Buses, 191

Car parking, 164
Cesar Pelli, 109
Chambre de Commerce et d'Industrie de Lyon, 82
Channel Tunnel, 10, 171
Channel Tunnel Rail Link (CTRL), 142
Channel Tunnel Terminal, Cheriton, Kent, 131–2, *131–2*
Charles de Gaulle Airport, *see* Paris Charles de Gaulle Airport
Chartered Institute of Logistics and Transport (CILT), 36
Chicago O'Hare Airport, 55–61, *55–61*
Circular Quay Interchange, Sydney, 125–6, *125–6*
City of Yokohama Port & Harbour Bureau Construction Department, 157
Civil Aviation Authority data (UK traffic figures), 38
Coach hub, 16, *17–18*
Commercial:
   motives, 192
   opportunities, *22–9*, 163
Community hub, *25–7*
Constraints on airport building form, 174
Converging standards, 8
CPMG Architects, 87
CRB Architectes, Lyon, 82
Crest Nicholson Properties, 192
Croydon Airport, 173
CUSS (Common-user self-service), 179
Customs, 183
CUTE (Common-user terminal equipment), 179

Dense mixed use, *28–9*
Disability Discrimination Act, 163
DMJM, 114
Docklands Light Railway (DLR), 136
Doppelmayr, 86, *91*
Düsseldorf Airport interchange, 118–22

EGS Design, 134
Enschede Station, 146–8, *147–8*
Euro Airport (Basel–Mulhouse), 39

European Commission's Trans European Networks fund, 87
European rail-air integration, 9, 191
European Social Fund, 146
Eurostar, 127, 142, 171
Eurotunnel plc, 131–2

Faber Maunsell, 134
Farnborough railway station, *193*
Fentress & Bradburn, 114
First Group, 191
Foreign Office Architects Ltd, 157
Foster & Partners:
   Hong Kong, 76
   Luton, 117
   St Pancras, 142
   Stansted, 73
Frankfurt Airport, 65–9, *65–9*
Fraport AG, 65
Friends of the Earth, 15
Fritsch, Chiari & Partner ZT GmbH, Vienna, 46
'Future Development of Air Transport in the UK: South East', 8

Gatwick, *see* London Gatwick
Gatwick Express, 74
Gillet, Guillaume, 82
Grand Junction Link, 13–15, *13–14*
Groupe Air France, 82
Guildford railway station, *192*

Halliburton KBR, 16
Hamburg–Harburg Technical University, 39
Heathrow, *see* London Heathrow
Heathrow Express, 48, 51, 103, 142
Helen Hamlyn Research Centre at the Royal College of Art, London, 19
Hewdon Consulting, 16
Highways Agency, UK, 9, 16
Hinchman and Grylls, 108
Hoar, Frank, 4
Hochtief, 120
Hong Kong Airport, 76–8, *77–8*
Hong Kong Airport Authority, 76

IAA (Enschede), 148
Immigration, 181, *181*
Inchon Airport, Seoul, South Korea, 113–15, *113–15*
Information, *31–3*
Information Technology, 192
Integrated transport, 7–15

195

Index

International Air Rail Organisation (IARO), 34
International Air Transport Association (IATA), 34, 163, 177
International Civil Aviation Organisation (ICAO), 186
Itten and Brechbuhl AG, 38

Jackaman, Morris, development of London Gatwick, 3
Jacobs, formerly Gibb, 128
JSK International Architekten und Ingenieure GmbH, 65, 120
Jubilee Line, 136, 192

Korean Architects Collaborative, 114

Lawrence Nield & Partners Australia Pty Ltd, 158
Leo Daly, 109
Levels of service, 162, 163
Light rail systems, 172
Loading bridges, 184, *185*
London Gatwick, 74–5, *75*
 location, 9
 original terminal, 3, *4–5*, 173
London Heathrow:
 access, 15
 Central Terminal Area, 103–5, *103–5*
 constraints, 9
 hub, 10
 satellite airport, 10
 Terminal 4, 51–4, *51–4*
 Terminal 5, 48–50, *48–50*
London Oxford, 10–13, *10–12*
London Stansted, 9, 70–3, *70–3*
London Underground Ltd, 141
Luton Airport, 9
 interchange, 116–17, *116–17*
Lyon Perrache Railway Station, 123–4, *123–4*
Lyon St Exupéry Airport, 9, 82–5, *83–5*

Maglev (Magnetic Levitation), 86
Manchester Airport, 92–102, *92–102*
Manchester Airport Developments Ltd, 96
Manchester Airport plc, 98
Manchester Metrolink, 92, 133, *172*
Manchester Piccadilly Station, 133–5, *133–5*
MARTA, Atlanta, 108
Mass Transit Railway, Hong Kong, 76
Metropolitan Washington Airport Authority, 109
Metrorail, Washington, 109
Minority Airport Architects and Planners, 108
Mott MacDonald, 76
Multiple railway-station/bus and coach/car interchanges:
 contiguous, 127–53

 vertical separation, 123–6
Murphy & Jahn, 55

NACO, Netherlands Airport Consultants, 45
National Center for Intermodal Transportation, USA (NCIT), 37
National Exhibition Centre, 86–7
Network Rail, 134, 193
Nicholas Grimshaw & Partners, 38
Nick Derbyshire Design Associates, 128
No frills airlines, 8, 9, 116
Nottingham University – database, 37
NSW Public Works Department and the Maritime Services Board, 158

Obstacle clearances, 174
Oscar Faber & Partners (now Faber Maunsell), 98

P&O, 154
Paris Charles de Gaulle Airport, 62–4, *62–4*
Parkway stations, 9, *16*
Pascall & Watson Architects, 73
Passenger:
 kilometres per year, 34
 segregation in airport terminals, 175
Passengers' luggage in advance, 8
Perkins & Will, 57, 118
Perth Station, Western Australia, 152–3, *152–3*
Peter Brett Associates, 106
Piccadilly Line, 48, 51, 103
Planning procedures, 9
Pleiade Associates, *10–12*
Portland International Airport, 79–81, *79–81*
Price & Myers,148

Rail interface, 171–2, *171–2*
Rail Link Engineering (RLE), 142
Rail station forecourts, 16, *19*
Railway builders, 8
Reclaiming the interchange (research programme), 19–33, 193
Residual spaces, *30–1*
Richard Rogers Partnership, 50
Ronald Reagan Washington National Airport, 109–10, *109–10*
Rotterdam Central Station, 149–51, *149–51*
Royal Institute of Chartered Surveyors (RICS), 36

Samoo Architects and Engineers, 114
San Francisco International Airport, 111–12, *111–12*
Santiago Calatrava, 85
Schiphol Airport, *see* Amsterdam Schiphol Airport

Scott Brownrigg Ltd:
 Farnborough, 193
 Guildford, 192
 Heathrow Terminal 4, 51
 Lyon, 82
 Manchester Airport, 98
 partnership with RCA, 19
Scott Wilson, 4, 51, 98
Security, 8, 162, 179, *181*
Ship and ferry terminals, 154–61
Southampton Airport, 106–7, *106–7*
Southampton Mayflower Terminal, 154–6, *155–6*
Space standards, 162
Spending patterns, 191
St Pancras Station, 142–5, *143–5*
Standard Busy Rate (SBR), 173
Stansted, *see* London Stansted
Stansted Express, 70, *72*
Stevens and Wilkinson/Smith, 108
STOLports, 174
Strategic Rail Authority, 15
Stratford Station, London, 136–41, *136–41*
Swiss Ernst Basler & Partners, 38
Sydney Overseas Passenger Terminal, 158–61, *158–61*

Technip TPS, Lyon, 82
Terry Farrell & Partners, 114
Thameslink trains, 117, 142
The Manser Practice, 106, 154
Thyssen Henschel, *42*
Trams, 191
Transport Development Areas (TDAs), 36
Transport for London, 37
Tri-met, Portland, *80–1*

Underground Transport Interchanges, 16–19, *16–19*
Union Railways (part of London and Continental Railways), 142
Unique (Flughafen Zurich AG), 38
Urban:
 connector, *22–3*
 ribbons, *23–4*
URS Thorburn Colquhoun, 134

Vienna Airport, 46–7, *46–7*

Wilkinson Eyre, 141

Yokohama port terminal, Japan, 157, *157*
YRM Architects Designers Planners, 16, 74

Zurich Airport, 39–42, *39–42*